T0133511

AI and SWARM

Evolutionary Approach to Emergent Intelligence

Hitoshi Iba

Information and Communication Engineering
School of Information Science and Technology
The University of Tokyo, Japan

CRC Press
Taylor & Francis Group
Boca Raton   London   New York

CRC Press is an imprint of the
Taylor & Francis Group, an **informa** business

A SCIENCE PUBLISHERS BOOK

CRC Press
Taylor & Francis Group
6000 Broken Sound Parkway NW, Suite 300
Boca Raton, FL 33487-2742

Printed on acid-free paper
Version Date: 20190725

International Standard Book Number-13: 978-0-367-13631-4 (Hardback)

Library of Congress Cataloging-in-Publication Data

Names: Iba, Hitoshi, author.
Title: AI and swarm : evolutionary approach to emergent intelligence /
    Hitoshi Iba, Information and Communication Engineering, School of
    Information Science and Technology, The University of Tokyo, Tokyo,
    Japan.
Description: First edition. | Boca Raton, FL : CRC Press/Taylor & Francis
    Group, 2019. | Includes bibliographical references and indexes. |
    Summary: "The book provides an overview of swarm intelligence and
    adaptive complex systems for AI, organized in seven chapters. Each
    chapter begins with a background of the problem and of the current state
    of the art of the field, and ends with detailed discussion about the
    complex system frameworks. In addition, the simulators based on
    optimizers such as PSO, ABC complex adaptive system simulation are
    described in detail. These simulators, along with the source codes that
    are available online, will be useful to readers for practical
    application tasks"-- Provided by publisher.
Identifiers: LCCN 2019029264 | ISBN 9780367136314 (hardback ; acid-free
    paper)
Subjects: LCSH: Artificial intelligence. | Swarm intelligence. |
    Evolutionary computation. | Self-organizing systems.
Classification: LCC Q335 .I333 2019 | DDC 006.3--dc23
LC record available at https://lccn.loc.gov/2019029264

Visit the Taylor & Francis Web site at
http://www.taylorandfrancis.com

and the CRC Press Web site at
http://www.crcpress.com

Preface

We used to joke that AI means "Almost Implemented."
—Rodney Brooks

This book is an explanatory text about Artificial Intelligence (AI) and swarm, that is, "emergence" for AI. These days, we can say that we are in the middle of the third AI boom. One of the factors responsible for this is deep learning, which resulted in the evolution of neural networks as well as machine learning based on statistical theory. However, there remains the question of whether it is possible to create a true AI (a human-like general-purpose intelligence, the so-called "strong" AI). Many research results indicate that there is clearly more depth in human intelligence and cognition than is possible to express by AI techniques ("weak" AI).

For about 30 years, the author has studied evolution and emergent computation. The swarm mechanism gives us a glimpse of the principles of intelligence in humans and living beings. In this book, such topics related to artificial intelligence are explained in detail, from the theoretical background to the most recent progress, as well as future topics.

Some readers may wonder if some of the topics related to swarm do really belong to the realm of AI technology. Indeed, the essence of AI is to find problems. According to this line of thought, a problem is no longer considered to be AI as soon as it is solved. Brooks' joke in the beginning of this text symbolizes this point. Brooks is a researcher specializing in AI at MIT (Massachusetts Institute of Technology) since the 1980's. He is the author of numerous innovative ideas in the field of AI. He is also famous for having created the robot vacuum cleaner Roomba (he is the founder of *iRobot* in the US). From this perspective, is it possible to say that, because the vacuum cleaner robot is already in the market, it is no longer AI?

Current, AI seems to be surrounded by an excessively flamboyant advertisement of its technical applications. Of course, it is worth noting that the intention here is not to deny all this. The author himself is involved in such projects. However, being satisfied with the above is a different issue. Given that it is possible to link any type of intelligence to AI, we should not ignore basic research that focuses on the essence of cognition and life.

The objective of the author's research is to build a model of human cognitive functions from the perspective of "emergence" in order to achieve strong AI. In other words, the objective is to understand how intelligence appeared and interacted with the real world, as well as the causal relationships in terms of actual physical and chemical mechanisms. In that sense, this book intends not only to explain intelligent behavior itself, but also understand its causes and effects. In order to do that, I have explained the emergence of several types of human cognitive errors, cognitive dissonances, irrational behaviors, and cooperative/betrayal actions.

Unfortunately, it is difficult to explain such phenomena using machine learning based on statistical models and deep learning based on big data, which constitute the core of current AI. On the contrary, a few hypotheses that contradict such approaches may be needed. In this book, research results related to brain mechanism are described, which is a factor that directly triggers intelligent behavior (called "proximate cause" or "physiological cause" in biological jargon), and the reasons for the evolution of living beings' behavior ("ultimate cause," or "biological/evolutional cause"). A few emergence simulation models are also introduced for better understanding. These approaches are important research efforts to clarify the essence of intelligence, in order to achieve true AI.

Most photographs depicting nature (modes of life and shapes of fish, patterns found in animals, images of Galapagos, Papua New Guinea, etc.) were shot by the author himself. Unfortunately, it is difficult for us to witness the evolution of nature on our own. However, it is possible to feel the driving force of emergence and the significance of diversity. In order to do that, it is suggested that you leave behind your PC or smartphone and see real things with your own eyes. The author has indulged in water life observation for about 30 years as a hobby, and still feels moved by it, reinforcing his conviction that nothing compares to seeing the real thing. Even though it is now possible to obtain almost any kind of image from the Internet, there are lots of things to be learned from real nature, and they are worth seeing.

Numerous things can be learned from actual living beings and from nature when studying swarms and AI. Therefore, in this book, the author has tried as much as possible to explain topics related to living beings and real life based on actual examples. It is the intent of the author that the readers will expand those topics and tackle new issues related to AI and swarm.

Acknowledgments

This book is based on my class notes on "Artificial Intelligence" and "Simulation Theory". The assignments given to the students in class were often unusual and considered difficult or unanswerable. However, the reports that were submitted in return often contained impressive descriptions and deep insights which I always appreciated. In this book, some examples of answers and opinions contained in the submitted reports are used after necessary corrections and refinement. Although it is impossible to acknowledge everyone by their names, I would like to express his deepest gratitude to the students who made the efforts to write interesting reports.

Undoubtedly, those philosophical and interesting discussions with colleagues from the laboratory and institute where I belonged as a student, constitute the core of this book. I takes this opportunity to express my deepest gratitude to teachers, and senior and junior colleagues.

I also wish to express my gratitude to SmartRams Co., Ltd. for cover image design.

And last, but not least, I would like to thank his wife Yumiko and sons and daughter Kohki, Hirono and Hiroto, for their patience and assistance.

H. Iba

April 2019   Amami isl. (Galapagos of the East), Japan

Contents

Abbreviations

ABC	:	artificial bee colony
ACO	:	ant colony optimization
AI	:	artificial intelligence
AL	:	artificial life
ANN	:	artificial neural network
ASEP	:	asymmetric simple exclusion process
BCA	:	Burgers cellular automaton automata
CA	:	cellular automata
CS	:	cuckoo search
CSO	:	cat swarm optimization
DLA	:	diffusion-limited aggregation
EA	:	evolutionary algorithms
EANN	:	evolutionary ANN
EC	:	evolutionary computation
EDA	:	estimation of distribution algorithm
EDD	:	earliest due date
EP	:	evolutionary programming
ES	:	evolution strategy
ESS	:	evolutionarily stable strategy
FA	:	firefly algorithm
FIFO	:	first-in, first-out
GA	:	genetic algorithms
GP	:	genetic programming
HS	:	harmony search
IPD	:	iterated prisoner's dilemma
JSSP	:	job scheduling problem
OBP	:	occlusion-based pushing
PSO	:	particle swarm optimization

SA   : simulated annealing
SPT   : shortest processing time
TFT   : tit for tat
TSP   : traveling salesman problem
UCB   : upper confidence bound
WSLS   : Win-Stay-Lose-Shift
ZD   : zero-determinant strategy

Chapter 1

Introduction

A sonnet written by a machine would be better appreciated by another machine.

—Alan Turing

1.1 What is AI? – Strong AI vs Weak AI

AI (artificial intelligence) refers to the implementation of intelligence on a computer and is divided into two hypotheses.

> **Strong AI** The viewpoint that true intelligence can be implemented on a computer, also known as general AI.
>
> **Weak AI** The viewpoint that computers can merely give the impression of intelligence, also known as narrow AI.

In the same manner, artificial life (AL), which is mentioned later, can be defined in terms of strong AL and weak AL.

"Weak AI" is an application of AI technologies to enable a high-functioning system that replicates human intelligence for a specific purpose. In fact, there are breakthroughs in weak AI.

In this book we consider simulation in the sense of "strong AI". More precisely, the rationale behind this approach is that "the appropriately programed computer really is a mind, in the sense that computers, given the right programs, can be literally said to understand and have other cognitive states."

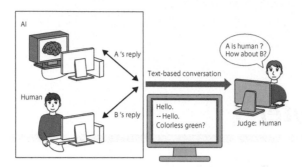

Figure 1.1: Turing test.

Realizing such a computer is nontrivial, since the definition of "intelligence" is difficult in the first place. Therefore, if a claim is made that AI (in the strong sense) has been created, what would be the most appropriate way to evaluate it?

To this end, Alan Turing[1] proposed a test of a machine's capacity to exhibit intelligent behavior, now called the "Turing test", which, despite being powerful, is the subject of numerous disputes (Fig. 1.1). The question of whether machines can think was considered in great depth by Turing, and his final opinion was affirmative. The Turing test can be translated into modern terms in the form of a game involving the exchanging of messages via a discussion board:

■ One day, two new users, A and B, join the discussion board.

■ When a message is sent to A and B, they both return apt responses.

■ Of A and B, one is human and the other is a computer.

■ However, it is impossible to determine which is which, regardless of the questions asked.

If a program passes this test (in other words, the computer cannot be identified), the program can be said to simulate intelligence (as long as the questions are valid). A similar contest, named the "The Loebner prize" after its patron, the American philanthropist Hugh Loebner, is held online[2]. Although an award of 100,000 US dollars and a solid gold medal has been offered since 1990, so far, not a single machine participating in the contest has satisfied the criteria for winning.

Nevertheless, a number of problems with the Turing test have been pointed out, and various critical remarks have been issued about potential implementation

---

[1] Alan Turing (1912–1954): British mathematician. He worked on deciphering the Enigma encryption used by the German Army during World War II. He is considered the "father of computer science." The Turing machine for computing theory and the Turing model for morphogenesis are some of his pioneering achievements. Apple's bitten apple logo is allegedly an homage to Turing.

[2] http://www.loebner.net/Prizef/loebner-prize.html

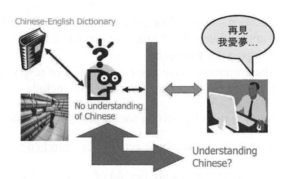

Figure 1.2: A Chinese room.

of AI. A notable example is the challenge to the very "definition of intelligence" by John Searle[3], who questioned the foundations of the Turing test by creating a counter thought experiment. Searle's experiment, known as the "Chinese room," can be summarized as follows.

A person is confined to a room with a large amount of written material on the Chinese language (Fig. 1.2). Looking inside the room is impossible, and there are only input and output boxes for submitting sentences and obtaining responses. Having no understanding of Chinese, the person cannot distinguish between different Chinese characters (for this purpose, we assume that the person is British and not Japanese). Furthermore, the person is equipped with a comprehensive manual (written in English) containing rules for connecting sets of Chinese characters. Let us consider that a person who understands Chinese is leading a conversation by inserting questions written in Chinese into the input box and retrieving answers from the output box. Searle provides the following argument.

Suppose that the person quickly becomes truly proficient at manipulating Chinese characters in accordance to the instructions, and that the person outside the room also becomes proficient at providing instructions. Then, the answers prepared by the person inside the room would become indistinguishable from answers provided by a Chinese person. Nobody would consider, simply by looking at the provided answers, that the person inside the room does not understand Chinese. However, in contrast to English, in the case of Chinese, the person in the room prepares the answers by formally manipulating characters, without any understanding whatsoever.

It cannot be said that true understanding is achieved simply by looking at the typeface while performing manipulations in accordance with a formal set of rules. However, as demonstrated by the "Chinese room" thought experiment,

---

[3]John Rogers Searle (1932–): American philosopher. He is known for his criticism of artificial intelligence. In 1980, he proposed the "strong and weak AI" classification.

under specific conditions, human-like behavior can be fabricated by both humans and machines if appropriate formal rules are provided. Searle, therefore, argues that strong AI is impossible to realize.

Various counterarguments have been considered in response to Searle, and questions that would probably occur to most people include

- Can conversion rules be written for all possible inputs?

- Can such an immense database actually be searched?

However, these counterarguments are devoid of meaning. The former rejects the realization of AI in the first place, and the latter cannot be refuted in light of the possibility that ultra-high-speed parallel computing or quantum computing may exist in the future. Thus, neither one can serve as the basis of an argument.

One powerful counterargument is based on system theory. Although the person in the room certainly lacks understanding, he constitutes no more than a single part of a larger system incorporating other elements, such as the paper and the database, and this system as a whole does possess understanding. It is the equivalent to the fact that, even for Chinese, cranial nerve cells do not understand the language by themselves. This point is integral to the complex systems regarded in this book. The level at which intelligence is sought depends on the observed phenomenon, and if the phenomenon is considered as being an emergent property, the validity of the above system theory can be recognized. Moreover, a debate is ongoing about whether intelligence should be thought of as an integrated concept or as a phenomenon that is co-evolving as a result of evolution.

On the other hand, Searle reargues that "if the manual is fully memorized and the answers are given without external help, it is still possible not to understand the Chinese language". Recently, Levesque et al. [67] counterpoint against Searle's criticism has been made based on a theory of computation. This is described in the "addition room argument" as follows:

- Consider the addition room, where twenty ten-digit numbers are to be added.

- Suppose there is a person in the room who does not know how to add and a manual on addition is placed in the middle of the room.

- At this time, if the human perfectly grasps the content of the manual and does all the operations in their head, then is it still possible to create a manual without understanding addition?

For the "addition room," the following manual immediately comes to mind:

- Memorize one-digit additions.

- Reduce two or more-digit numbers to one-digit and add.

However, this is exactly the way we learned in elementary school. Therefore, knowing this manual means being acquainted with the addition algorithm, thus, it is possible to say that addition is understood.

Let us think about a little bit more primitive manual described below:

■ Go to the same chapter as the first number.

■ Go to the section with the same number as the second number within that chapter.

■ Furthermore, go to the subsection with the same number as the third number within that section.

■ Repeat that for all twenty numbers.

■ When everything is over, there is a number with up to twelve-digits, so return that number and exit.

This manual merely lists the calculation results in the same way as a dictionary. Just like in the Chinese room argument, a human following this manual does not add and additionally does not understand the calculation at all. So, was Searle's claim correct?

Let us consider here the computational complexity of this manual. The chapter corresponding to the first number requires ten to the tenth power. Each chapter consists of ten to tenth power of sections. Since this is repeated twenty times, it becomes 10 to 10 power to 20 power $= 10^{200}$ of items. It is said that the number of molecules in the universe is around 10^{100}. Therefore, such a large manual can never be made. From this, it can be understood that Searle's claim is mistaken, from a computational theory. This is true as far as this manual is concerned, but Searle himself did not specify the composition method of the manual, so there may be recurring objections.

Recently, machine translation is accomplished using a huge database and statistical processing. The famous one is Google's machine translation. Past machine translation was a classical AI based on natural language understanding. Unfortunately, it was not always effective. On the other hand, Google's machine translation does not understand natural language at all. This software statistically joins translated words using a huge database (human translations, e.g., minutes of the United Nations). As a result, it won the 2006 machine learning contest with an overwhelming difference. This kind of translation method is possible due to the power of modern computers and internet connection. This approach may be a solution to the Chinese room argument.

1.2 What is Emergence?

Originally, emergence (emergent property) was a term in biology. A good example of that property is seen in ant social life. Each ant has a simple mechanical behavior. However, the ant colony as a whole acts in a highly intelligent collective manner (collective behavior), according to the distribution pattern of food and enemies, therefore increasing the survival rate of the colony. As a result, a caste system emerges, consisting of individuals specialized in doing different jobs, where a social division of labor and cooperation will be seen (see Fig. 1.3). There are no rules (programs) that govern the behavior of the whole colony. Nonetheless, an emergence of intellectual global behavior, brought about by collective action of simple programs (individual ants), is observed.

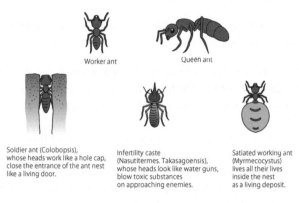

Worker ant

Queen ant

Soldier ant (Colobopsis), whose heads work like a hole cap, close the entrance of the ant nest like a living door.

Infertility caste (Nasutitermes. Takasagoensis), whose heads look like water guns, blow toxic substances on approaching enemies.

Satiated working ant (Myrmecocystus) lives all their lives inside the nest as a living deposit.

Figure 1.3: Job seperation emerged [123].

In emergence, arising of the macroscopic phenomena from micro interactions has become a central idea. This phenomenon is observed in various fields, such as morphological formation (pattern generation, see sec. 5.3) or group formation (see sec. 3.2) by animal populations in ecology. Emergent phenomenon can be seen even in economy and finance. For example, in a financial market, each individual (participants, investors) pursues their own profit based on local information, thus, interacts with other individuals (speculative behavior). As a result, when we look at the entire financial market, phenomena such as "price movement of sidelines" and "nervous movement" are observed. On the other hand, each market participant does not necessarily behave in a nervous manner or wait watchfully. In this manner, generation of a macro phenomenon that is not described by micro interactions is emergence. E.O. Wilson[4] claims that "the higher nature of life is emergent [109]". In other words, the higher-level phenomenon, leading to the creation of biological individual from cells and society from in-

---

[4]Edward Osborn Wilson (1929–): American entomologist. Researcher of sociobiology and biodiversity. See also the footnote of 188 page.

dividuals, cannot be described only by knowledge of the nature of components constituting lower levels.

There are two types of emergence [32]:

Weak emergence New property appears as a result of the interactions of elemental levels. Emergent property can be reduced to individual components.

Strong emergence The newly emerged property cannot be reduced because it is more than the sum of its components. Laws governing the property cannot be predicted by understanding the laws governing the structure of another level.

By becoming a group, it spontaneously acquires properties and trends not included in the underlying rule. Emergence is recognized in various fields, such as physics, biology, chemistry, sociology, and art. However, strong emergence cannot be dealt with in traditional physics and reductionist approaches.

The author thinks that human intelligence could be explained through emergence. However, many brain researchers and neuroscientists are critical towards emergence. Many of them take a reductionist method and dislike the idea of a ghost in the brain, which cannot be explained deterministically. On the other hand, it is known that the combination of the network configuration to derive the same behavior is extremely large. Therefore, it is a significant obstacle for the neuroscientist, because further analysis of the neural circuit leads only to an understanding of the mechanism, and not how it actually operates. Recognizing that there are different levels in the organization is crucial to the understanding of the phenomenon of emergence [32]. Recently, the mechanism by which directives traverse from micro to macro level in a causal manner is being mathematically researched by the neuroscientists [50].

1.3   Cellular Automaton and Edge of Chaos

The eminent mathematician John von Neumann[5] studied self-reproducing automata in 1946, shortly before his death. He found that self-reproduction is possible with 29 cell states, and proved that a machine could not only reproduce

---

[5]Jon von Neumann (1903–1957): Hungarian-born American mathematician. As symbolized by the gossip that circulated around Princeton that he slept with a mystery novel as his pillow in the day and learned mathematics from the devil at night, he conducted pioneering research in various areas, such as mathematics, physics, engineering, computer science, economics, meteorology, psychology, and political science, having substantial influence in posterior advances in these areas. He was also deeply involved with American nuclear policy and was part of the Manhattan project.

Table 1.1: State of cell in next step.

Current state of cell	States of neighbor cells	State in next step
On	two or three are "on"	On
	Other cases	Off
Off	three are "on"	Off
	Other cases	Off

itself, but could also build machines more complex than itself. The research stopped because of his death; however, in 1966, Arthur Burks edited and published von Neumann's manuscripts. John Conway, a British mathematician, expanded on the work of von Neumann and, in 1970, introduced the Game of Life, which attracted immense interest. Some people became "Game of Life hackers," programmers and designers more interested in operating computers than in eating; they were not the criminal hackers of today. Hackers at MIT rigorously researched the Game of Life, and their results contributed to advances in computer science and artificial intelligence. The concept of the Game of Life evolved into the "cellular automata" (CA), which is still widely studied in the field of artificial life. Most of the research on artificial life shares much in common with the world where hackers played in the early days of computers.

The Game of Life is played on a grid of equal-sized squares (cells). Each cell can be either "on" or "off". There are eight adjacent cells to each cell in a two-dimensional grid (above and below, left and right, four diagonals). This is called the Moore neighborhood. The state in the next step is determined by the rules outlined in Table 1.1. The "on" state corresponds to a "•" in the cell, whereas the "off" state corresponds to a blank. The following interesting patterns can be observed with these rules.

1. Disappearing pattern (diagonal triplet)

2. Stable pattern (2×2 block)

   • •   • •   • •
   • • ⇒ • • ⇒ • • ⇒ remains stable

3. Two-state switch (Flicker, where vertical triplets and horizontal triplets appear in turn)

   •
   • ⇒ • • • ⇒ • ⇒ repeats
   •   •

4. Glider (pattern moving in one direction)

"Eaters" that stop gliders and "glider guns" that shoot gliders can be defined, and glider guns are generated by the collision of gliders. Such self-organizing capabilities mean that the Game of Life can be used to configure a universal Turing machine. The fundamental logic gates (AND, OR, NOT) consist of glider rows and disappearing reactions, and blocks of stable patterns are used as memory. However, the number of cells needed in a self-organizing system is estimated to be about 10 trillion (3 million×3 million). The size would be a square whose sides are 3 km long, if 1 mm^2 cells are used.

Consider a one-dimensional Game of Life, one of the simplest cellular automata. The sequence of cells in one dimension at time t is expressed as follows:

$$a_t^1, a_t^2, a_t^3, \cdots \tag{1.1}$$

Here, each variable is either 0 (off) or 1 (on). The general rule used to determine the state a_{t+1}^i of cell i at time $t+1$ can be written as a function F of the state at time t as

$$a_{t+1}^i = F(a_t^{i-r}, a_t^{i-r+1}, \cdots, a_t^i, \cdots, a_t^{i+r-1}, a_t^{i+r}) \tag{1.2}$$

Here, r is the radius, or range of cells that affects this cell.

For instance, a rule for $r = 1$,

$$a_{t+1}^i = a_t^{i-1} + a_t^i + a_t^{i+1} \pmod 2 \tag{1.3}$$

results in the determination of the next state as follows:

```
time t    :   0010011010101100
time t+1  :   *11111001010001*
```

An interesting problem is the task of finding the majority rule. The task is to find a rule that would ultimately end in a sequence of all 1 (0) if the majority of the cells are 1 (0) with the minimum radius (r) possible for a one-dimensional binary sequence of a given length. The general solution to this problem is not known.

A famous example is a majority rule problem with length 149 and radius 3. The problem is reduced to finding a function that assigns 1 or 0 to an input with 7 bits ($= 3 + 1 + 3$, radius of 3 plus itself); therefore, the function space is 2^{2^7}.

How can a cellular automaton (CA) obtain a solution to the majority problem?

One method is to change the color (black or white) of a cell to the majority of its neighboring cells. However, this method does not work well, as shown in Fig. 1.4, because it results in a fixed pattern divided into black and white.

In 1978, Gacs, Kurdyumov and Levin found the rules (GLK) regarding this problem. Lawrence Davis obtained an improved version of these rules in 1995, and Rajarshi Das proposed another modification. There is also research to find effective rules through GAs (genetic algorithms) or GP (genetic programming). The concept of Boolean functions is applied when GP is used. The fitness value is defined by the percentage of correctly processed sequences out of 1000 randomly generated sequences of length 149.

Rules determined by various methods are summarized in Table 1.2. Here, the transition rules are shown from 0000000 to 1111111 in 128-bit form. In other words, if the first bit is 0,

$$F(000\,0\,000) = 0 \tag{1.4}$$

Table 1.3 is a comparison of different rules. The rules obtained using GP were very effective; reference [4] contains the details of this work.

Figure 1.5 shows how a CA obtained by GA can solve this problem well [75, 76]. The regions that were initially dominated by black or white cells become regions that are completely occupied by either black or white cells. A vertical line always exists at locations where a black region to the right meets a white region to the right. In contrast, a triangular region with a chessboard pattern forms where a white region to the right meets a black region to the right.

Figure 1.4: CA carrying out majority voting (Oxford University Press, Inc., [76]).

The two edges of the growing triangular region at the center with a chessboard pattern grow at the same pace, progressing the same distance per unit time. The left edge extends until it collides with a vertical boundary. The right edge barely avoids the vertical boundary at the left (note that the right and left edges are connected). Therefore, the left edge can extend for a shorter length, which means that the length of the white region limited by the left edge is shorter than

Table 1.2: Majority rules.

Name of rule (year)	Transition rules
GKL(1978)	00000000 01011111 00000000 01011111 00000000 01011111 00000000 01011111 00000000 01011111 11111111 01011111 00000000 01011111 11111111 01011111
Davis(1995)	00000000 00101111 00000011 01011111 00000000 00011111 11001111 00011111 00000000 00101111 11111100 01011111 00000000 00011111 11111111 00011111
Das(1995)	00000111 00000000 00000111 11111111 00001111 00000000 00001111 11111111 00001111 00000000 00000111 11111111 00001111 00110001 00001111 11111111
GP(1995)	00000101 00000000 01010101 00000101 00000101 00000000 01010101 00000101 01010101 11111111 01010101 11111111 01010101 11111111 01010101 11111111

Table 1.3: Performance in the majority problem.

Rule	Performance	Number of tests
GKL	81.6%	10^6
Davis	81.8%	10^6
Das	82.178%	10^7
GA	76.9%	10^6
GP	82.326%	10^7

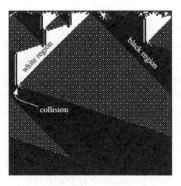

Figure 1.5: Behavior of CA driven by GA (Oxford University Press, Inc., [76]).

the length of the black region limited by the right edge. The left edge disappears at the collision point, allowing the black region to grow. Furthermore, the two edges disappear at the bottom vertex and the entire lattice row becomes black, showing that the correct answer was obtained.

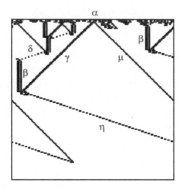

Figure 1.6: Explanation of CA behavior from collision of particles (Oxford University Press, Inc., [76]).

Melanie Mitchell analyzed the information processing structure on CA that evolved through GA by using the behavior of dynamic systems [75, 76]. The boundaries between simple regions (edges and vertical boundaries) are considered carriers of information, and information is processed when these boundaries collide. Figure 1.6 shows only the boundaries in Fig. 1.5. These boundary lines are called "particles" (similar to elementary particles in a cloud chamber used in physics). The particles are represented by Greek letters following the tradition in physics. Six particles are generated in this CA. Each particle represents a different type of boundary. For instance, η is the boundary between a black region and a chessboard-patterned region. A number of collisions of particles can be observed. For example, $\beta + \gamma$ results in the generation of a new particle η, and both particles annihilate in $\mu + \eta$.

It is easy to understand how information is coded and calculated when the behavior of CA is expressed in the language of particles. For instance, α and β particles are coded with different information on the initial configuration. γ particles contain information that this is the boundary with a white region, and a μ particle is a boundary with a white region. When a γ particle collides with a β particle before colliding with a μ particle, this means that the information carried by β and γ particles becomes integrated, showing that the initial large white region is smaller than the initial large black region that shares a boundary. This is coded into the newly generated η particle.

Stephen Wolfram[6] systematically studied the patterns that form when different rules (eq. (1.2)) are used. He grouped the patterns generated by one-dimensional CA into four classes.

---

[6]Stephen Wolfram (1959–): British theoretical physicist. At the age of 20, he obtained his Ph.D. degree from California Institute of Technology for his research on theoretical physics. In 1987, he found the company Wolfram Research. He released the mathematical software Mathematica, which is used for research in natural sciences and numerous other fields.

Class I All cells become the same state and the initial patterns disappear. For example, all cells become black or all cells become white.

Class II The patterns converge into a striped pattern that does not change or a pattern that periodically repeats.

Class III Aperiodic, chaotic patterns appear.

Class IV Complex behavior is observed, such as disappearing patterns or aperiodic and periodic patterns.

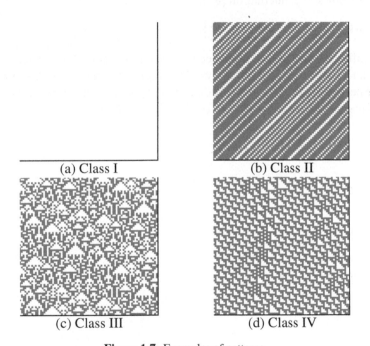

(a) Class I   (b) Class II

(c) Class III   (d) Class IV

Figure 1.7: Examples of patterns.

Examples of these patterns are shown in Fig. 1.7. The following are the rules behind these patterns (radius 1):

■ Class I: Rule 0

■ Class II: Rule 245

■ Class III: Rule 90

■ Class IV: Rule 110

Here, transition rules are expressed in 2^{2^3} bits from 000 to 111, and the number of the rule is the decimal equivalent of those bits. For instance, the transition rules for Rule 110 in Class IV are as follows [71]:

$$01101110_{binary} = 2 + 2^2 + 2^3 + 2^5 + 2^6 = 110_{decimal} \qquad (1.5)$$

In other words, 000, 100 and 111 become 0, and 001, 010, 011, 101, 110 become 1.

Rule 110 has the following interesting characteristics in computer science.

1. It is computationally-universal [126].

2. It shows $1/f$ fluctuation [83].

3. Its prediction is P-complete [82].

Kauffman and Packard proposed the concept of the "edge of chaos" from the behavior of CA as discussed above. This concept represents Class IV patterns where periodic patterns and aperiodic, chaotic patterns are repeated. The working hypothesis in artificial life is "life on the edge of chaos."

Chapter 2

AI, Alife and Emergent Computation

It is intriguing that computer scientists use the terms genotype and phenotype when talking about their programs.

—John Maynard Smith, [111]

2.1 Evolutionary Computation

2.1.1 What is Evolutionary Computation?

EAs (Evolutionary Algorithms) and EC (Evolutionary Computation) are methods that apply the mechanism of biological evolution to problems in computer science or in engineering. EAs consider the adaptation process of organisms to their environments as a learning process. Organisms evolve over a long period of time by repeating the evolution process whereby species that do not adapt to the environment become extinct and species that adapt thrive. This can be applied to practical problems by replacing "environment" with "problem" or "information that was learned" and "fitness" with "goodness of the solution."

The most important operators in EAs are selection and genetic operations. Evolution of individuals does not happen if individuals that adapt better to their environments are surviving but nothing else is happening. Mutation and crossover by sexual reproduction result in the generation of a diverse range of individuals, which in turn promotes evolution. Selection is a procedure where good individuals are selected from a population. Species that adapt better to the environment are more likely to survive in nature. The selection procedure artificially carries out this process.

The typical examples of EAs are Genetic algorithms (GA) and Genetic programming (GP). They are the basic mechanisms for simulating complex systems. Next sections describe these methods in detail with practical applications.

GAs have the following characteristics:

■ Candidate solutions are represented by sequences of characters.

■ Mutation and crossover are used to generate solutions of the next generation.

Elements that constitute GAs include data representation (genotype or phenotype), selection, crossover, mutation, and alternation of generation. The performance of a search is significantly influenced by how these elements are implemented, as discussed below.

The data structure in GAs is either genotype (GTYPE) or phenotype (PTYPE). The GTYPE structure corresponds to chromosomes of organisms (see Fig. 2.1), and is a sequence representing a candidate solution (for example, a bit sequence having a fixed length). This structure is subject to genetic operations such as crossover and mutation. The implementer can design how to convert candidate solutions into sequences. For instance, a GTYPE structure can be obtained by conversion of a candidate solution into a sequence of integers that is then concatenated. On the other hand, PTYPE structures correspond to organisms, and are candidate solutions obtained by interpreting GTYPE structures. The fitness values of candidate solutions are calculated for PTYPE structures.

The GA is based on the concept of Darwinian evolution, where individuals who adapt better to their environments leave more offspring, while less fit individuals are eliminated. Individuals that adapt to their environments are candidate solutions that are better solutions to a problem, and the measure is the fitness of PTYPE structures.

The following selection methods have been proposed. In particular, the tournament selection is frequently used because scaling is not necessary. In all methods, individuals that have higher fitness are more likely to be selected.

■ Roulette selection
The roulette selection selects individuals with a probability in proportion to their fitness. This is the most general method in EAs, however, procedures such as scaling are necessary to perform searches efficiently[1].

■ Tournament selection
The tournament selection is widely used in EAs. This is a strategy in which a certain number (called as tournament size) of individuals are

---

[1] See **Steps** 5 and 6 on page 5 for detailed example.

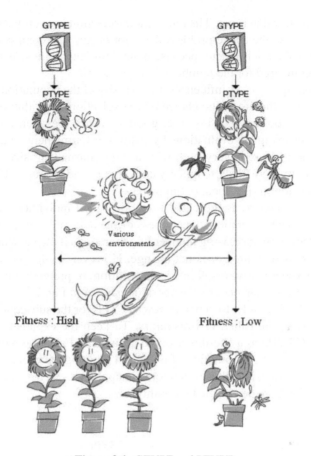

Figure 2.1: GTYPE and PTYPE.

randomly selected from a population, and the best of these individuals is finally chosen. This process is repeated until the size of the population is reached.

■ Truncate selection
Individuals are sorted based on fitness and the top $P_s \times M$ individuals are selected in truncate selection, where M is the population size and P_s is the selection rate. The selection pressure is very high, therefore, this method is not used in standard GP, but is often used in the estimation of distribution algorithm (EDA), which is an expansion of the GA. The computation cost of this method besides the cost for sorting is very low.

The selection rate in roulette selection is determined by the absolute value of the fitness. However, the selection pressure may become too high or too low with roulette selection in problems where the hierarchy of fitness is important but the absolute value is not. The tournament selection uses only the hierarchy of fitness, therefore, the above problem does not occur. The computational cost of tournament selection is greater because many individuals are selected and fitness values are compared for the number of tournaments.

Selection significantly influences the diversity of the population and the speed of convergence, therefore, the choice of the selection algorithm and its parameters is very important. For instance, good solutions are often observed with a small number of fitness evaluations by using the tournament selection because the selection pressure is very high with a large tournament size. However, the calculations are more likely to quickly converge to and be trapped in an inappropriate local solution (i.e., local optimum).

A different strategy is the elitist selection (good individuals are always included in the next generation). The fitness of the best individual never decreases in this strategy, with increasing number of generations if the environment against which fitness is measured does not change. However, using the elitist selection too early in a search may result in a local solution, or premature convergence.

When reproduction occurs, the operators shown in Fig. 2.2 are applied to the selected GTYPE in order to generate new GTYPE for the subsequent generation. These operators are called GA operators. To keep the explanation simple, we express the GTYPE as a one-dimensional array here. Each operator is analogous to the recombination or mutation of a gene in a biological organism. Generally, the frequency with which these operators are applied, as well as the sites at which they are applied, are determined randomly.

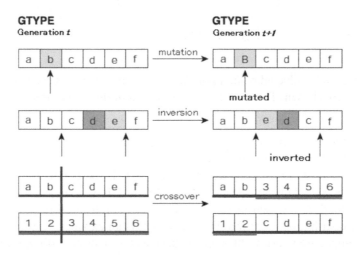

Figure 2.2: GA operators.

Crossover is an analogy of sexual reproduction where new offspring are generated by combining two parent individuals. There are a number of crossover methods based on the level of granularity in separating each individual, for example, the one-point crossover and the uniform crossover.

The crossover shown in Fig. 2.2 has one crossover point, so it is called a one-point crossover. Below are some methods for performing the crossover operation:

1. One-point crossover

2. Multi-point crossover (n-point crossover)

3. Uniform crossover

We have already explained the one-point crossover operation (Fig. 2.3(a)). The n-point crossover operation has n crossover points, so if $n = 1$, this is equivalent to the one-point crossover operation. With this crossover method, genes are carried over from one parent alternately between crossover points. A case in which $n = 3$ is shown in Fig. 2.3(b). Two-point crossovers, in which $n = 2$, are often used. Uniform crossovers are a crossover method in which any desired number of crossover points can be identified, so these are realized using a mask for a bit string consisting of 0 or 1. First, let us randomly generate a character string of 0s and 1s for this mask. The crossover is carried out as follows. Suppose the two selected parents are designated as Parent A and Parent B, and the offspring to be created are designated as Child A and Child B. At this point, the genes for offspring Child A are carried over from Parent A when the corresponding mask is 1, and are carried over from Parent B when the mask is 0. Conversely, the genes for offspring Child B are carried over from Parent A when the corresponding mask is 0, and are carried over from Parent B when the mask is 1 (Fig. 2.3(c)).

Mutation in organisms is considered to happen by mutation of nucleotide bases in genes. The GA mimics mutation in organisms by changing the value of a gene location (for example, changing 0 to 1 or 1 to 0). Mutation corresponds to errors during copying of genes in nature, and is implemented in GAs by changing a character in an individual after crossover (inversion of 0 and 1 in a bit sequence, for instance). Using crossover generally only results in a search of combinations of the existing solutions, therefore, using mutation to destroy part of the original gene is expected to increase the diversity of the population and, hence, widen the scope of the search. The reciprocal of the length of GTYPE structures is often used as the mutation rate, meaning that one bit per GTYPE structure mutates on average. Increasing the mutation rate results in an increase of diversity, but with the tradeoff of a higher probability of destroying good partial solutions.

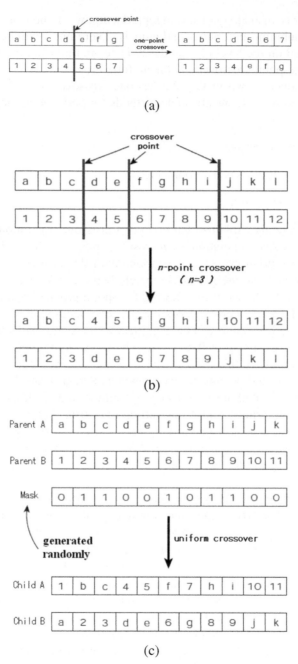

Figure 2.3: GA crossovers: (a) one-point crossover, (b) *n*-point crossover and (c) uniform crossover.

The flow of a GA is as follows:

Step1 Randomly generate sequences (GTYPE) of the initial population.

Step2 Convert GTYPE into PTYPE and calculate fitness for all individuals.

Step3 Select parents according to the selection strategy.

Step4 Generate individuals in the next generation (offspring) using genetic operators.

Step5 Check against convergence criteria and go back to **Step**2 if not converged.

Replacing parent individuals with offspring generated through operations such as selection, crossover, and mutation to obtain the population of the next generation is called the alternation of generation. Iterations finish when an individual with an acceptable fitness is found or a predetermined number of generations have been produced. It is also possible to continue the calculations for as long as possible if resources are available, and to terminate when sufficient convergence is achieved or further improvement of fitness appears difficult.

All individuals in the first generation are randomly generated individuals in EAs. EAs use genetic operators and therefore have less dependence on the initial state in comparison to the hill-climbing method, but an extremely biased initial population will decrease the performance of searches. Hence, the initial individuals must be generated to distribute in the search space as uniformly as possible. Figure 2.4 shows an example of a good initialization and a bad initialization. If the triangle △ in Fig. 2.4 is the optimal solution, the optimal solution is expected to be found in a relatively short time from the initial state in Fig. 2.4(a) because the initial individuals are relatively randomly distributed over the search space. On the other hand, the initial distribution in Fig. 2.4(b) has fewer individuals near the optimal solution and more individuals near the local optimal solution, therefore, it is more likely that the search will be trapped at the local optimal solution and it will be difficult to reach the optimal solution.

The initial individuals in GAs are determined by randomly assigning 0 or 1 in each gene location, and as the lengths of genes in GAs are fixed, this simple procedure can result in a relatively uniform distribution of initial individuals over the solution space.

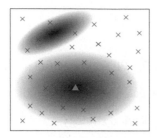 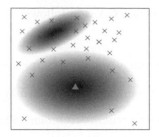

(a) Random initialization   (b) Nonuniform initialization

Figure 2.4: Examples of good and bad initialization. The triangles △ indicate the optimal solution.

2.1.2 *Evolution Strategy*

In Europe (especially in Germany), research groups with similar methodologies to genetic algorithm (GA) have been active for a long time. This method has been called Evolution Strategy (referred to as "ES," hereinafter), with Ingo Reichenberg as the central figure. In the initial stage, ES differed from GA in the following two points [74]:

1. It mainly uses mutation as an operator.

2. It deals with real-valued representation.

Each individual of ES is represented as a set of real number vectors, such as $(\vec{x}, \vec{\sigma})$, where $\vec{x}$ represents a position vector in a space of search while $\vec{\sigma}$ represents a standard deviation vector. The mutation is represented as follows:

$$\vec{x}^{t+1} = \vec{x}^{t} + N(\vec{0}, \vec{\sigma}), \tag{2.1}$$

where $N(\vec{0}, \vec{\sigma})$ is a random number created according to a Gaussian distribution of the average $\vec{0}$ (i.e., zero vector) and the standard deviation $\vec{\sigma}$.

At an early stage of ES, a search is performed using a group composed of single individuals. In this case, the offspring that results from a mutation ($\vec{x}^{t+1}$ in eq. (2.1)) is adopted as a member of the new group only if the adaptability is improved compared to that of its parent ($\vec{x}^{t}$). That is, it will become a parent for the next generation.

As opposed to GA, ES is not affected by crossover. Therefore, a quantitative investigation is not difficult and has resulted in the achievement of robust mathematical analysis related to different phenomena, such as the effect of the mutation rate [97, 108]. For instance, the convergence theorem has been proved [9]. Furthermore, **The $\frac{1}{5}$ rule** had been proposed to optimize the convergent rate.

The $\frac{1}{5}$ Rule
If the successful mutation ratio[2] is larger (smaller) than $\frac{1}{5}$, increase (decrease) $\overrightarrow{\sigma}$.

Actually, by observing the successful mutation ratio of the past k generation $\varphi(k)$, it was determined that the following equation controls the mutation:

$$
\overrightarrow{\sigma}^{t+1} = \begin{cases} c_d \times \overrightarrow{\sigma}^t, & if \ \varphi(k) < 1/5, \\ c_i \times \overrightarrow{\sigma}^t, & if \ \varphi(k) > 1/5, \\ \overrightarrow{\sigma}^t, & if \ \varphi(k) = 1/5. \end{cases} \tag{2.2}
$$

Particularly in reference [108], $c_d = 0.82$ and $c_i = 1/0.82$ were adopted. The intuitive meaning of this rule is that "If it brings success, continue to search with longer steps. Otherwise, shrink them."

Eventually, ES has been expanded as a search method using a group which consists of plural individuals. In this case, a crossover operator, average operator (calculates the mean of two-parent vectors), and other kinds of vector manipulation have been adopted in addition to the previously described mutation operator. Furthermore, what makes it different from GA is that ES takes the following two types of methods as the selection method:

1. $(\mu + \lambda) - ES$
   A parent group comprised of μ individuals generates a child of λ individuals. Select μ individuals as parents of the next generation among the groups of the total $(\mu + \lambda)$ individuals.

2. $(\mu, \lambda) - ES$
   A parent group comprised of μ individuals generates a child of λ individuals (on condition that $\mu < \lambda$). Select μ individuals as a parent of the next generation among groups of λ individuals.

Generally speaking, $(\mu, \lambda) - ES$ is considered to be superior in an environment which changes with the progression of time, and when noise is present.

ES has, thus, been applied to various optimization problems [70], and to those other than real values [47]. Refer to the reference materials [74, 10] for additional information on ES and its comparison with GA.

2.1.3   *Multi-objective Optimization*

In life, we often wish to pursue two or more mutually conflicting goals (Fig. 2.5). In finance, return and risk are generally conflicting objectives. In assets management and investment, return is the profit gained through management and risk is the uncertainty and amplitude of fluctuations. Risk can be gauged from price volatility (the coefficient of variation). High volatility means large price fluctuations. Ordinary volatility is generally expressed in terms of standard deviation.

In general, risk and return are in a "trade-off" relationship. High-return, low-risk financial products would be ideal, but things are not actually that easy. Usually, the choice is one of the following two:

■ Low risk, low return : The possibility of loss of the principal is low, but management returns will be small.

■ High risk, high return: Losses may occur, but any returns might be large.

In considering the optimization problem, so far, we have limited the discussion to maximization (or minimization) of a single objective (i.e., the fitness function). As in risk versus return, however, it is often necessary to consider optimization when there are multiple objectives, that is, "multi-objective optimization."

Let us begin with an example [36]. Assume you are engaged in transport planning for a town. The means of reducing traffic accidents range from installing traffic lights, placing more traffic signs, and regulating traffic, to setting up checkpoints (Fig. 2.6). Each involves a different cost, and the number of traffic accidents will vary with the chosen approach. Let us assume that five means (A, B, C, D, and E) are available, and that the cost and the predicted accident numbers are:

$$A = (2,10)$$
$$B = (4,6)$$
$$C = (8,4)$$
$$D = (9,5)$$
$$E = (7,8),$$

where the first element is the cost and the second is the predicted accident number, as plotted in Fig. 2.6. The natural impulse is to desire attainment of both goals in full: The lowest cost and the lowest predicted accident number. Unfortunately, it is not necessarily possible to attain both objectives by the same means and, thus, not possible to optimize both at the same time.

In such situations, the concept of "Pareto optimality" is useful. For a given developmental event to represent a Pareto optimal solution, it must be the case

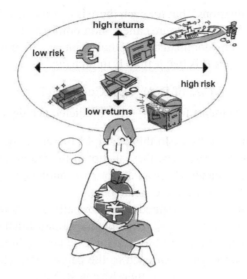

Figure 2.5: Risk vs returns.

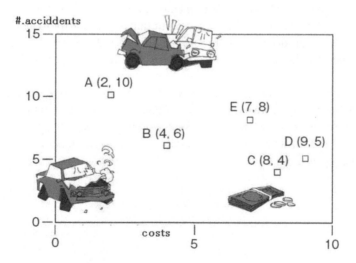

Figure 2.6: Cost vs expected numbers of accidents.

that no other developmental events exist which are of equal or greater desirability, for all evaluation functions, that is, fitness functions.

Let us look again at Fig. 2.6. Note that the points in the graph increase in desirability as we move toward the lower left. A, B, and C in particular appear to be good candidates. None of these three candidates is the best in both dimensions, that is, in both "evaluations," but for each there is no other candidate that is better in both evaluations. Such points are called "non-dominated" points. Points D and E, in contrast, are both "dominated" by other points and therefore less desirable. E is dominated by B, as B is better than E in both evaluations:

$$\text{Cost of } B(4) \quad < \quad \text{Cost of } E(7)$$
$$\text{Predicted accidents for } B(6) \quad < \quad \text{Predicted accidents for } E(8)$$

D is similarly dominated by C. In this example, therefore, the Pareto optimums are A, B, and C. As this suggests, the concept of the Pareto optimum cannot be used to select just one candidate from a group of candidates, and, thus, it cannot be concluded which of A, B, and C is the best.

Pareto optimality may be defined more formally as follows. Let two points $x = (x_1, \cdots, x_n)$ and $y = (y_1, \cdots, y_n)$ exist in an n-dimensional search space, with each dimension representing an objective (an evaluation) function, and with the objective being a minimization of each to the degree possible. The domination of y by x (written as $x <_p y$) may be therefore be defined as

$$x <_p y \iff (\forall i)(x_i \leq y_i) \wedge (\exists i)(x_i < y_i). \tag{2.3}$$

In the following, we will refer to n (the number of different evaluation functions) as the "dimension number." Any point that is not inferior to any other point will be called "non-dominated" or "non-inferior," and the curve (or curved surface) formed by the set of Pareto optimal solutions will be called the "Pareto front."

On this basis, it is possible to apply GA to multi-objective optimization. In the following example, we use the VEGA (Vector Evaluated Genetic Algorithms) system of Schaffer et al. [105, 106]. Selection with VEGA is performed as follows (see Fig. 2.7), where n is the evaluation dimension number (the number of different evaluation functions).

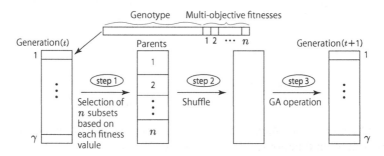

Figure 2.7: VEGA (vector evaluated Genetic Algorithms) algorithm.

1. Let the n subsets be Sub_i $(i = 1, \cdots, n)$.

2. In Sub_i, retain the individual selected only by evaluation function i.

3. Mix $Sub_1, Sub_2, \cdots, Sub_n$ and shuffle.

4. Produce the next-generation offspring, using the genetic operator on these sets.

Note that the selection is based upon every evaluation function (i.e., the function value in every dimension). In contrast, reproduction is performed not for each subset but rather for the complete set. In other words, crossover is performed for the individuals in Sub_i and $Sub_j(i \neq j)$. In this way, we preserve the superior individuals in each dimension and at the same time select the individuals that are superior to the average in one or more dimensions.

Let us now consider the VEGA operation, as applied to optimization of the binary function:

$$\begin{aligned} F_{21}(t) &= t^2, \\ F_{22}(t) &= (t-2)^2, \end{aligned}$$

where t is the only independent variable. The graphical appearance of this function is shown in Fig. 2.8 (which is called as a "Pareto map"). The VEGA objective is to select non-dominated points, such as those shown in the figure.

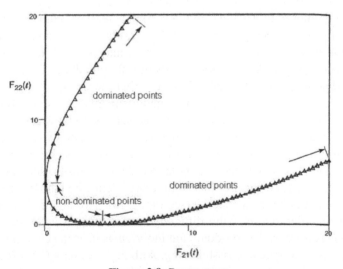

Figure 2.8: Pareto maps.

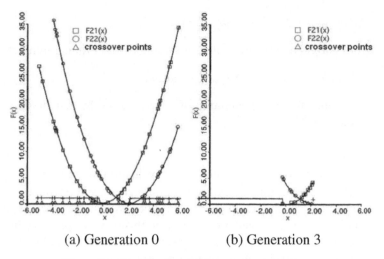

(a) Generation 0   (b) Generation 3

Figure 2.9: Results of multi-objective optimization.

Figure 2.9 shows the results for generations 0 and 3. The population size is set to 30 individuals in each dimension, and a crossover ratio of 0.95 and a mutation ratio of 0.01 is applied. As shown, the VEGA can effectively find a front (a Pareto front) containing no dominated points. However, it has lost several intermediate points.

VEGA selection, as may be seen in this example, involves a problem that may be described as follows. The selection pressure (in biological terms, the degree of selectivity applied to the survival and proliferation of the organism's population) on the optimum values (A and C in Fig. 2.6) in at least one dimension (evaluation function) works as desired. In the unlikely event that a utopian individual (an individual superior in all dimensions) exists, it may be possible to find it through genetic operation on the superior parents in only one dimension. In most cases, however, no utopian individual exists. It is then necessary to obtain the Pareto optimal points, some of which are intermediate in all dimensions (B in Fig. 2.6).

In an ideal GA, it is desirable to have the same selection pressure applied on all of these, and, thus, in Fig. 2.6 on all three of A, B, and C. With VEGA selection, however, an intermediate point such as B cannot survive. In effect, species differentiation occurs within the population, with dimension-specific superiority of each species. The danger of this is more prevalent when the Pareto optimal region is convex rather than concave (see Fig. 2.10). Two modifications have been proposed in order to resolve this difficulty. One is to apply an excess of heuristic selection pressure on the non-dominant individuals in each generation. The other is to increase interspecies crossbreeding. Although ordinary GA selection is random, this is effective because utopian individuals are more readily produced by interspecies than intraspecies' crossover.

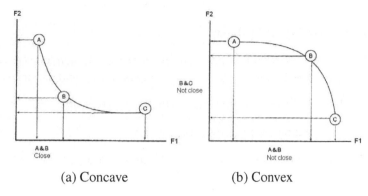

| (a) Concave | (b) Convex |

Figure 2.10: Pareto fronts.

2.2 How to Make a Bit? – Exploration vs Exploitation

In this section, let us consider how to deal with gambling. This issue is concerned not only with gambling, but it is also very important in artificial intelligence and economics.

So, in the first place, what is gambling? One of the simplest bets, which we will take into consideration, is the two-armed bandit machine[3]. Let us consider a slot machine with two levers (R, L) as is presented in Fig. 2.11. This problem is defined as follows:

Figure 2.11: A slot machine with two levers.

---

[3]In English, it is referred to as two-lever slot machine problem. The name of the problem comes from the fact that slot machine gambling can wind up player's money (extort money from player).

Two-Armed Bandit Problem

1. The payoff is paid according to the different payment rate (payoff rate) for R and L. Each of these has averages μ_R, μ_L, and variances σ_R^2, σ_L^2.

2. It is not known beforehand whether $\mu_R > \mu_L$ or $\mu_R < \mu_L$

3. The total number of bets is set to be N.

Under these conditions, what is the best way to bet on arms R and L?

The multi-armed bandit problem is defined similarly, as M-lever slot machine ($M \geq 0$).

Now, it is necessary to consider two phases:

Exploration Determine if $\mu_R > \mu_L$ or $\mu_R < \mu_L$.

Exploitation According to the exploration, bet on the lever with the better payoff rate.

This problem falls into a dilemma because N is finite. Too much exploration (investigation) leads to a local search, which is weak to noise and will be more difficult to adapt to the error (variance) of the payoff rate. On the other hand, if it is exploited too much, information gained from exploration will be ignored and the chance to earn money will be lost.

In recent years, the UCT algorithm has been proposed, which is popularly used for Monte Carlo tree search in games. In this algorithm, the following UCB (Upper Confidence Bound) value is employed for the approximate solution of multi-armed bandit problem in order to decide on promising action:

$$
\text{UCB value} \ = \ \frac{\text{Number of times when the lever of interest won}}{\text{Number of playouts allocated to the lever of interest}}
$$
$$
+ \ c \sqrt{\frac{2\log\left(\text{Number of playouts}\right)}{\text{Number of playouts allocated to the lever of interest}}}
$$

Where c is a user defined constant. Second term of this equation is a play-out[4] and the smaller it becomes, the bigger the term grows. On the other hand, the first

---

[4]Randomly choosing a legal hand to play the game and to end it.

term becomes smaller with the worsening of the action. Therefore, if the number of play-outs is small and unluckily the game is lost, the evaluation value is decreased, which, in fact, avoids the problem of not playing the game even though there are promising actions left. UCB value can be thought of as evaluation value expressing trade-off between exploration (the second term of 2.4) and exploitation (the first term). In other words, it evaluates which levers are good and which ones are bad, at the same time trying to use good levers (i.e., levers with better payoff rates) for more play-outs. By using the UCB value, it is possible to make an optimal selection under specific conditions[5].

Let us actually experiment with the slot machine. Here, we observe the total payoff money (i.e., reward) acquired by selecting and pulling each lever of the two-lever and five-lever slot machines, assuming that 1,000 coins are given. However, winnings will not be used again (that is, it ends with 1,000 trials). Let us vary the average and variance of the payoff money of each lever. Let $i = 1, 2, \cdots$ denote levers, and the total payoff for each lever i is $Q_n(i)$ for the slot turned n times (after choosing one lever). n_i is the number of times lever i has been pulled. Of course, $n_1 + n_2 + \cdots = n$. Let $r_j(i)$ be a payoff for pulling i-th lever for j-th time (note that $0 \leq j \leq n_i$). Also, from the above definition, $Q_n(i) = \sum_{j=1}^{n_i} r_j(i)$ holds.

The strategy to compare is as follows:

1. Choose lever randomly

2. Greedy method: Choose the lever i with the highest average payoff $\frac{Q_n(i)}{n}$ to pull next.

3. ε–Greedy method: Same as greedy method, but lever is chosen randomly with the probability ε. In this experiment, probability was chosen to be $\varepsilon = 0.01$.

4. UCB1 method: The following is used as the UCB value. However, n_i is the number of times lever i has been pulled so far.

$$\text{UCB value} \quad = \quad \frac{Q_n(i)}{n_i} + \sqrt{\frac{2 \log n}{n_i}}$$

5. Revised UCB1 method: The following is used as the UCB value.

$$\text{UCB correction value} \quad = \quad \frac{Q_n(i)}{n_i} + \sqrt{\frac{c \log n}{n_i}}$$

---

[5]In n trials, the loss generated from choosing the best lever for each time is suppressed to the value $O(\log(n))$. However, constraints such as the payoff money being in the range of $[0, 1]$ are imposed. It has been shown that, under this condition, no algorithm can exceed the UCB value [6].

However,

$$c = \min\left[\frac{1}{4}, \hat{\sigma}_i^2(n) + \sqrt{\frac{2\log n}{n_i}}\right],$$

where $\hat{\sigma}_i^2(n)$ is an estimated variance. For example, as an unbiased estimate of the population variance, it is calculated as follows:

$$\hat{\sigma}_i^2(n) = \frac{1}{n_i - 1}\sum_{j=1}^{n_i}\left(r_j(i) - \frac{Q_n(i)}{n_i}\right)^2$$

This method practically has a better performance than the UCB1 method, but it does not have a theoretical guarantee.

Let us first consider when there are two levers. The average of each payoff is set to be 0.10, 0.05, and the variance for both payoffs is set to be 0.20. The payoffs are normally distributed with this mean and variance, according to Gaussian distribution (i.e., normal distribution). A typical example of execution is shown in Fig. 2.12. In this case, the total number of betting N (maximum number of trials) is set to 1,000. (a) is the total reward (i.e., the sum of payoffs) for each trial. Eventually, UCB1 and revised UCB1 are shown to give good results. On the other hand, greedy method and ε–greedy method are almost the same as a random selection. This indicates that the selection is confused by the variance value and is unable to choose the optimal lever (see Fig. 5.8(b)). It becomes more apparent if the number of levers is increased to 10. Figure 2.13 shows the experimental result with the average payoff of 10 levers being set to: 0.10 0.05 0.05 0.05 0.02 0.02 0.02 0.01 0.01 0.01, and variance of all payoffs is set to 0.20. In this case, only the first lever is shown to be optimal. As can be seen in Fig. 2.13, after exceeding 200 trials both UCB1 and revised UCB1 accurately select the optimal lever. In contrast, the greedy method achieves grades lower than random selection. Table 2.1 shows the average value after repeating the experiment 200 times. The best-performing method is highlighted in gray. The advantages of UCB1 and revised UCB1 are presented in the table.

On the other hand, UCB1 is not necessarily always good. Depending on conditions, greedy method can be superior. For example, Table 2.2 and Table 2.3 are experimental results with the average payoff and its variance when the number of levers is 2 and 5, respectively (number of bets $N = 1,000$, repetition number 50). In most cases, when the values of the variance and the average payoff are the same, results of greedy methods are improved. Also, greedy methods seem to find good results, if the average payoff value is biased and extremely good. If the average value is the same, then random and greedy methods work well in contrast with UCB1, which sometimes cannot fully learn the variance. However, there is not much of a difference in results in any of the methods.

For another type of payoff, consider the Bernoulli type (see section 4.2.1). This is a method in which the probability μ_i is defined for each lever i that pays 1.0 with probability μ_i and does not provide compensation with probability $1 - \mu_i$. This remuneration can be thought of as a click model of online advertisements. In this case, each lever becomes an advertisement, and a user browsing an advertisement is equal to experimenting with a slot machine. In addition, payoffs are equivalent to clicks, and it becomes the problem of selecting an advertisement maximizing the total clicks. The results are presented in Table 2.4. In such a situation, the advantage of ε–greedy method stands out. It is also worth noting that the results of a simple greedy method are not that good. UCB1 is superior to the simple greedy method, but results were not as good as for normal distribution type payoff. This seems to reflect that there is no variance in the winnings.

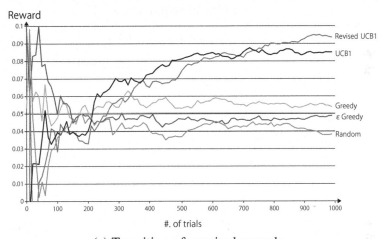

(a) Transition of acquired rewards.

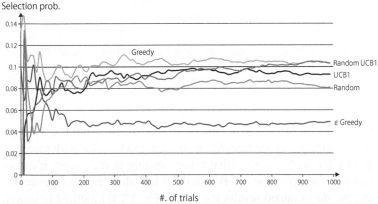

(b) Transition of probability to select the optimal lever (no. 1).

Figure 2.12: Two-lever slot machine. The payoffs are normally distributed.

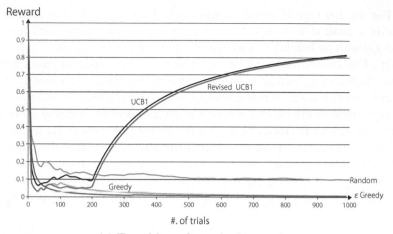

(a) Transition of acquired rewards.

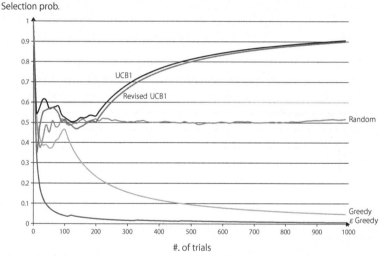

(b) Transition of probability to select the optimal lever (no. 1).

Figure 2.13: Ten-lever slot machine. The payoffs are normally distributed.

Let us summarize the above results. It is surprising that the greedy method, which is the simplest procedure, has achieved good results with many problems. Since the UCB1 method is sensitive to the variance of the payoff, it can cope with high variance payoffs. The UCB1 method has slower convergence than other methods, but the obtained results are better. The UCB1 method is superior when few levers or high variance payoffs are considered, however, results degrade as the number of levers increases. Also, when the payoff is of the normal distribu-

Table 2.1: Experimental results with normally distributed payoffs. Variance values are set to be the same. The total rewards are shown with different methods. The best-performing method is highlighted in gray.

#. of levers	2	10
Payoff average	0.10 0.05	0.10 0.05 0.05 0.05 0.02 0.02 0.02 0.01 0.01 0.01
Payoff variance	0.20	0.20
Random	75.335725	34.405275
Greedy method	91.970039	67.387339
ε–Greedy method	85.579438	73.442488
UCB1	94.317254	77.593354
Revised UCB1	95.036724	85.177174

Table 2.2: Experimental results with normally distributed payoffs. The number of levers is two. The total rewards are shown with different methods. The best-performing method is highlighted in gray.

Payoff average	0.0, 5.0	0.0, 5.0	0.0, 5.0	5.0, 5.0
Payoff variance	10.0, 40.0	40.0, 10.0	10.0, 10.0	40.0, 10.0
	High risk high return	Low risk high return	Same variance	Same average
Random	2382.234993	2466.161786	2501.014429	5033.966115
Greey method	3590.420261	4544.600735	4745.431955	4995.128383
ε–Greedy method	3098.290123	4783.067266	4744.443555	5015.799637
UCB1	4304.636240	4821.753778	4760.705372	5006.838874
Revised UCB1	4478.568354	4807.838191	4776.344278	4996.441438

Table 2.3: Experimental results with normally distributed payoffs. The number of levers is five. The total rewards are shown with different methods. The best-performing method is highlighted in gray.

Payoff average	0.0,2.0,2.0, 1.0,1.0	0.0,2.0,2.0, 1.0,6.0	1.0,1.0,1.0, 0,1.0	1.1,1.2,1.3, 1.4,1.5
Payoff variance	10.0,1.0,2.0 1.0,2.0	10.0,1.0,2.0 1.0,2.0	1.0,2.0,3.0 4.0,5.0	1.0,1.0,1.0 1.0,1.0
	Different variances	Different averages	Same average	Same variance
Random	1200.164188	2189.213874	999.646251	1299.583166
Greedy method	1828.578583	5595.396152	1002.104703	1367.964301
ε–Greedy method	1895.622982	1968.515276	1000.613158	1123.039962
UCB1	1930.462162	1919.132855	997.706195	1119.840872
Revised UCB1	1929.266064	1962.263051	999.876270	1118.288329

Table 2.4: Experimental results with payoffs according to Bernoulli distribution. Variance values are set to be the same. The total rewards are shown with different methods. The best-performing method is highlighted in gray.

#. of levers	2	10
Payoff average	0.10 0.05	0.10 0.05 0.05 0.05 0.02 0.02 0.02 0.01 0.01 0.01
Random	74.405	33.735
Greedy method	90.31	72.62
ε–Greedy method	100.39	100.1
UCB1	95.635	79.965
Revised UCB1	94.96	81.805

tion type, the results are better than for other distributions (e.g., Bernoulli type). This is because the best lever is well separated from the other levers in the normal distribution.

The slot machine problem has been extensively studied in the fields of statistical decision theory and adaptive control [11]. For example, John Holland[6] used the slot machine problem to create a mathematical model describing how genetic algorithms (GA) allocate individuals into a schema[7] [51]. This is one of the theoretical foundations of evolutionary computation. Holland proposed, through analytical analysis, one optimal strategy for two-lever slot machine problem. This method exponentially increases the probability of choosing a good lever compared with the probability to try out the wrong lever, based on an information obtained by trial. This is also true for schema sampling in GA. Holland's schema theorem claims, that the sub-optimal schema is sampled implicitly, resulting in an optimization of the online evaluation.

Evolutionary economics is a field that studies economics based on complex adaptive systems and explains economic activities from the perspective of evolutionary theories and biology. Keywords of this field are increasing returns and path dependency [5]. In economics, it refers to positive feedback, or so-called winning horse effect[8]. Economist Brain Arthur thought that the economy is a complex system operated by interactions of numerous agents. In conventional economics, it was common sense that the equilibrium between demand and sup-

---

[6]John Henry Holland (1929–2015): American scientist. He is known for his pioneering research on complex systems and nonlinear systems. In 1975, he published the epoch-making book "Adaptation in Natural and Artificial Systems" on genetic algorithms, which made him the father of that area. Holland's famous schema theorem is the basis of GAs.

[7]A partial structure maintained by a group of genes. Exploration in GA and GP proceeds by combining schemas by crossover.

[8]Also known as the bandwagon effect (see Fig. 2.14). A phenomenon that strengthens the support for the strategy through the information on the prevalence of the strategy. In the financial market, it points at a phenomenon of the exchange rates moving abnormally in one direction as a result of people piggybacking on the flow of the market.

Figure 2.14: Bandwagon effect.

ply was kept in balance by Adam Smith's "Invisible hand of God," and that the economy stabilizes accordingly[9]. However, it is clear that this trend does not hold if trends in recent stock prices and finance are analyzed. There is no equilibrium point, and a certain trigger can make it all (the whole market) to move in one direction, and it may not be possible to stop it. This is the increasing returns.

In addition, the path dependence of the technology selection means that the selection of a certain technology is determined by chance irrespective of the superiority or inferiority of the technology itself. Once the technology was chosen (decided on)[10], it cannot be changed anymore. It is so because, even if the drawbacks of the technology are revealed, the migration cost is too high.

One of the well-known examples is the QWERTY keyboard layout [22]. The top left row of the English notation keyboard we usually use is QWERTY. This is not optimal for typing English words. Why does it look like this?

A common belief is that the early stage of the typewriter (around the end of the 19th century) was designed to reduce the speed of typing to prevent the collision of arms. At that time the delicate machine arms broke by interfering when typing speed was too fast. Therefore, the frequently used E, A, and S were deliberately placed in positions that were hard to type.

In response to this, in the 1930s, August Dvorak from the University of Washington developed a DSK keyboard (the Dvorak Simplified Keyboard), that could be used to type more efficiently (Fig. 2.15(a)). As a more rational layout, more frequently used characters were placed around the home position[11] to eliminate excess finger movement. In fact, there was also a report that the DSK layout had a higher learning rate and more than 20% faster typing rate after mastering [87]. Figure 2.15(b) and (c) presents the awkward hand movement, taken from the video recording when actually typing a keyboard (the height of the figure shows

---

[9]Adam Smith (1723–1790): British economist. He organized the free market economy theory and is known as the father of classical economics.

[10]This is referred to as lock-in.

[11]Initial finger placement. Normally in the QWERTY arrangement, they are ASDF (left hand: From pinky finger towards the index finger) and JKL; (right hand: From index finger towards the pinky).

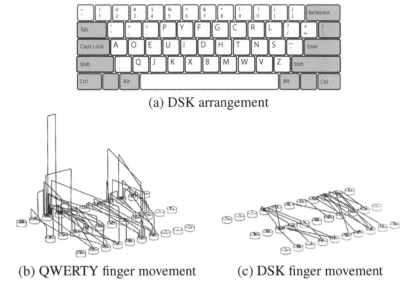

(a) DSK arrangement

(b) QWERTY finger movement   (c) DSK finger movement

Figure 2.15: Dvorak Simplified keyboard.

the frequency). The DSK layout is about one-tenth of the QWERTY arrangement.

However, since the present layout was proposed in 1882, QWERTY arrangement is the only one winning. Why is this? The reason that can be thought of, is that the QWERTY side trained typists by creating a typing school. As a result, many companies hired QWERTY trained typists. On the other hand, it is expensive to retrain typist to use DSK layout keyboards. This fact encouraged production of the QWERTY layout typewriters and gave birth to its users. The Apple II series, launched by Apple in 1984, came with a built-in switch to change between QWERTY and DSK layout keyboards, but it had no effect.

In this way, if one is chosen on the spur of the moment, even if it is not known whether it is really good or not, it may not be possible to stop anymore due to avalanche-like profit growth. The slot machine problem is used to mathematically analyze this kind of selection phenomena.

2.3 Wireworld: A Computer Implemented as a Cellular Automaton

Wireworld was invented by Brian Silverman in 1987 (the following description was based on the document [29]). This is a kind of cellular automaton, which is suitable for simulation of electronic logic circuits. Rules are simple, similarly to the game of life, but it is shown to be Turing complete. In the wireworld, the following four states are used for the cell:

■ Empty

■ Electron head

■ Electron tail

■ Conductor

"Electrons" are generated from two adjacent cells, electron tail and electron head. The transition rules of each of the states are as follows:

■ empty ⇒ empty

■ electron head ⇒ electron tail

■ electron tail ⇒ conductor

■ conductor ⇒ electron head if exactly one or two of the neighbouring cells are electron heads, otherwise remains conductor.

In the wireworld, a logic circuit can be constructed with these simple rules.

To draw a conductor in the cell space, place a conductor on the line and enclose it in a background (empty) cells. To let electrons flow through the conductor, replace the first cell of the conductor with electron tail and the second cell with electron head. Then, electrons will move along the conductor, according to transition rules defined above. In every step in the wireworld clock (one cycle), the conductor cell in front of the electron changes to the electron head cell. The electron head cell changes to electron tail, and the electron head cell returns to the conductor cell.

Figure 2.16 shows the operation of the cell diode. In the figure, empty cells are white (or are omitted). In this circuit, electrons are allowed to flow only in one direction. Note the behavior of the rectangle part of 2×3. Only the center cell on one side of the rectangle is the background cell, creating an empty space in the conductor. Electrons entering from the side without empty space are divided into two and pull the trigger of the conductor cell on the other side of the empty space. Electrons entering from the opposite side are divided into three, but the conductor cell surrounded by the three electron head cells is not ignited, thus, cannot propagate through the conducting wire. In the figure, it is shown that electrons flow to the right, but not to the left.

Figure 2.17 shows how to construct basic logic circuits (NOT, OR, AND, XOR). These circuits have electron inputs on the left, and the output conductor on the right. A loop-shaped conductor is placed at the input. Electrons (arrangement of the electronic tail and electronic head) placed into the loop of the input part

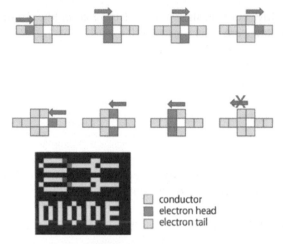

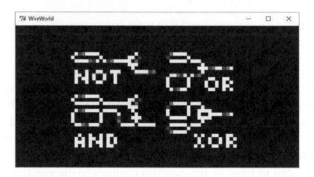

conductor
electron head
electron tail

Figure 2.16: Cell diode.

Figure 2.17: Basic logics (NOT, OR, AND, XOR).

continue to circle around in the loop. Electrons passing through the branching part are split into two and enter the input line of the logic. Note that the input frequency depends on the size of the loop.

Let us see the NOT operation (Fig. 2.18). Electron entering from the left loop reaches the NOT part on the right. Here, electrons are emitted from the top-right, and usually go out from the output on the bottom right. Note that, when there is a NOT input (electron from the left loop), it cancels out with top-right electron and nothing is output to the output line on the bottom-right. In this way, when the input is 1 the output is 0, and when the input is 0 the output becomes 1, thus, realizing NOT.

Further, in the OR circuit (Fig. 2.19), when electrons come from the left of one or both input the lines, they are sent out rightward. A diode-like circuit is necessary at the merging part so that the electrons coming from one input cannot flow back to another input.

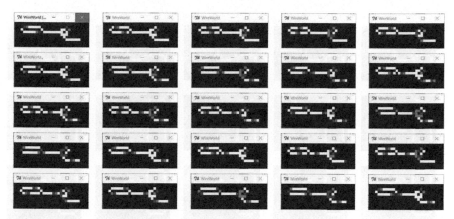

Figure 2.18: NOT operation.

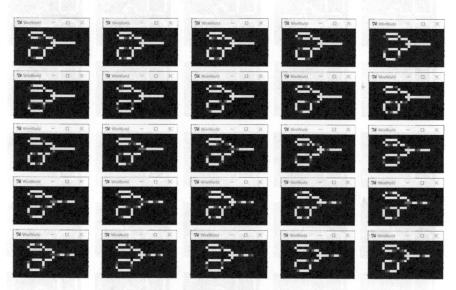

Figure 2.19: OR operation.

Color version at the end of the book

We also tried to simulate a circuit realizing a flip-flop (Fig. 2.20). This circuit is implemented by combining AND, NOT and OR. When a signal comes from the bottom input, the state in the flip-flop becomes 1, and the signal is output from the flip-flop every sixth cycle. When a signal comes from above, the state of the flip-flop is reset.

Furthermore, Fig. 2.21 shows a circuit realizing 3-bit binary adder. This circuit performs both input and output from the lower bit. In the figure, $(101)_2 = 5$ and $(110)_2 = 6$ are set as inputs to the top and bottom inputs, respectively. As

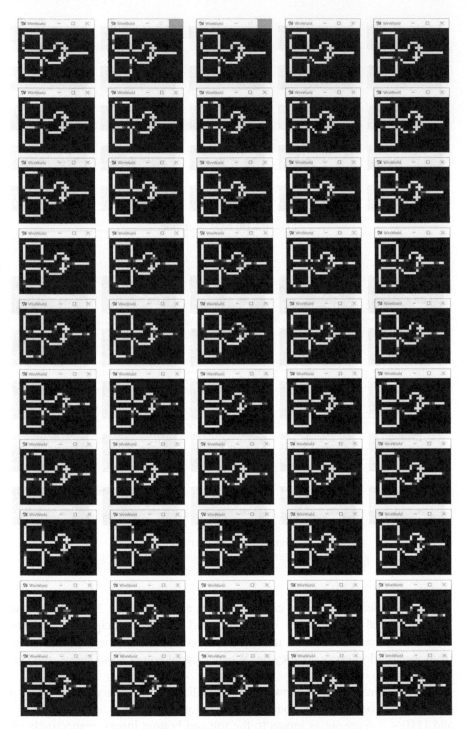

Figure 2.20: Flip Flop operation.

Figure 2.21: ADDER operation.

a result of execution, $(1011)_2 = 11$ was obtained as output, thus, it can be confirmed that addition is performed correctly.

2.4 Langton's Ant

Langton's ants is a puzzle invented by Christopher Langton[12], who is well known for his work in the area of artificial life.

---

[12]Christopher Langton (1949–): American computer scientist. In the late 1980's he coined the term Artificial Life and held the international conference "International Conference on the Synthesis and Simulation of Living Systems" (commonly known as Artificial Life I) at Los Alamos National Laboratory.

In this puzzle, squares on a plane (referred to as a "cell") are colored in either black or white. We then arbitrarily identify one square as the "ant." The ant can travel in any of the four cardinal directions at each step taken[13].

The movement of the "ant" is according to the following rules:

■ Upon moving onto a black square, turn 90° to the left.

■ Upon moving onto a white square, turn 90° to the right.

■ Change the color of the previous square of the ant from white to black or vice versa after moving to a new square.

Figure 2.22 shows the movement of an ant. In the figure, the ant faces to either the north, south, west or east. At the next time step, the ant travels according to the figure on the right. Since any color fits into the gray cells, the color should not be changed once it is determined.

Even with these simple rules, the ant demonstrates astonishingly complex movement. Figure 2.23 shows the ant's movement starting on a white cell in a torus lattice with dimensions of 80 by 80 (connecting bottom to up and right to left). A total of 10,184 steps formed a "highway" (regular pattern path) as shown in (a). Subsequently, a total of 15,094 steps (b), 34,476 steps (c) and 44,803 steps (d) also resulted in the formation of a highway[14]. After observing approximately 100,000 steps, the highway formation no longer existed.

Jim Propp discovered an interesting behavior of an ant in an indefinite space (not a torus) as summarized below:

■ The first few hundred of steps create beautiful symmetric patterns.

■ After an additional 10,000 steps, the continuously built path becomes considerably chaotic.

■ After a further 104 steps, a cycle whereby the ant crosses over two squares repeats indefinitely. Thus, a diagonal "highway" is established.

That is, the ants start the movement of creating a straight "highway" with no exception after wandering around for about 10,000 steps, although they show random movement at the beginning. This outcome does not depend on how the pattern was initially made. This implies that the "highway" is an attractor of Langton's ants.

Langston's ants can also be considered as a cellular automaton (see section 1.3). In this case, the background of the lattice is colored in black or white, and the ant has eight different color states based on the different combinations of current direction and background color.

---

[13] See section 3.1.1 for more precise simulation of ant colonies.
[14] Note that a highway bounces back due to the torus shaped space.

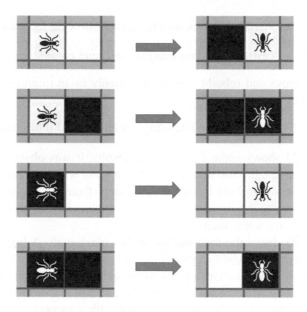

Figure 2.22: Movement of an ant.

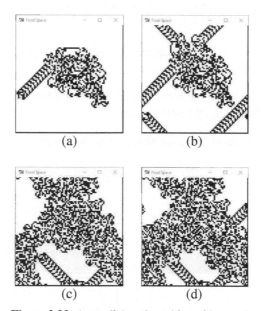

Figure 2.23: Ant trail (starting with a white map).

When the ant starts in a black rectangle, the following facts are observed:

- It builds "castle" including straight walls and complex projections.

- It destroys and rebuilds castles repeatedly in an interesting way as if it has a certain objective.

- It finally wanders around building a highway as if it has lost any interests.

Figure 2.24 shows some experimental results. It was observed that the ant initially built a "castle" and then a highway, starting with an initial state (a). The first highway has been built based on 9,337 steps from the beginning. Although it seems contrary to intuition, a highway is completed much faster compared to starting in only the white map (Fig. 2.23).

Figure 2.25 shows the case starting with a black wall. A highway was built with 5,035 steps (b), 10,968 steps and 28,992 steps, from the initial state (a).

When the start occurs with a black triangle, it has been observed that it cyclically moves along with the leg of a triangle, and builds a castle wall along with the base, as observed in the case of starting with a rectangle. A highway was built with 3,816 steps (b) and 7,797 steps (c) when starting with the initial state (a) shown in Fig. 2.26.

There are several unsolved problems associated with this puzzle (not a torus version), which are as follows:

- If the number of black squares is finite, is a highway built without fail irrespective of the initial pattern?

- Does an ant leave the area without fail after a long period of time elapses without being caught in the limited area?

- Does an ant leave the area along a highway without exception?

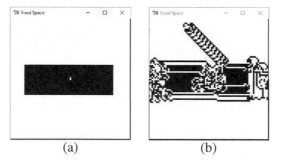

(a)   (b)

Figure 2.24: Ant trail (starting with a white map a black box).

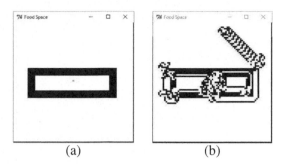

Figure 2.25: Ant trail (starting with a white map a black wall).

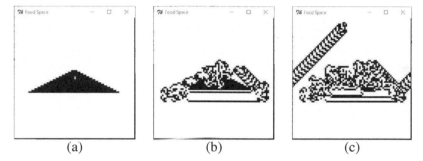

Figure 2.26: Ant trail (starting with a triangle).

Of the three proposed questions, affirmative evidence has been acquired with respect to the second. We shall examine these notions with various types of initial maps, and the remaining unsolved problems shall be investigated.

Chapter 3

Meta-heuristics

When you were a tadpole and I was a fish In the Paleozoic time,
And side by side on the ebbing tide We sprawled through the ooze and slime, ...

—Langdon Smith, 1858–1908, Evolution

3.1 Ant Colony Optimization (ACO)

3.1.1 Collective Behaviors of Ants

Ants march in a long line. There is food at one end, a nest at the other. This is a familiar scene in gardens and on roads, but the sophisticated distributed control by these small insects was recognized by humans only a few decades ago.

Ants established their life in groups, or colonies, more than a hundred million years before humans appeared on Earth. They formed a society that handles complex tasks such as food collection, nest building, and division of labor through primitive methods of communication. As a result, ants have a high level of fitness among species, and can adapt to harsh environments. New ideas, including routing, agents, and distributed control in robotics, have developed based on simple models of ant behavior. Applications of the ant behavior model have been used in many papers, and are becoming a field of research rather than a fad.

Marching is a cooperative ant behavior that can be explained by the pheromone trail model (Fig. 3.1). Cooperative behavior is frequently seen in ant colonies, and has attracted the interest of entomologists and behavioral scientists. Pheromones are volatile chemicals synthesized within the insect, and are used to communicate with other insects of the same species. Examples are sex pheromones that attract the opposite sex, alarm pheromones that alert group members, and trail pheromones that are used in ant marches.

Figure 3.1: Ant trail.

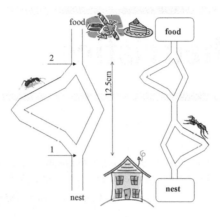

Figure 3.2: Bridge-shaped paths (adapted from Fig. 1a in [37], Springer-Verlag GmbH).

However, recent research indicates that pheromones are effective within a distance of only about $1m$ from the female. Therefore, it is still not known if males are attracted only because of the pheromones.

Many species of ants leave a trail of pheromones when carrying food to the nest. Ants follow the trails left by other ants when searching for food. Pheromones are volatile matter that is secreted while returning from the food source to the nest. Experiments shown in Fig. 3.2 by Deneubourg and Goss using Argentine ants linked this behavior to the search for the shortest path [37]. They connected bridge-shaped paths (two connected paths) between the nest and the food source, and counted the number of ants that used each path. This seems like a simple problem, but because ants are almost blind they have difficulty recognizing junctions, and cannot use complex methods to communicate the position of the food. Furthermore, all the ants must take the shorter path to increase the efficiency of the group. Ants handle this task by using pheromones to guide the other ants.

Figure 3.3 shows the ratio of ants that used the shorter path [37]. Almost every ant used the shorter path as time passed. Many of the ants return to the shorter path, secreting additional pheromones; therefore, the ants that followed also take the shorter path. This model can be applied to the search for the shortest path, and is used to solve the traveling salesman problem (TSP) and routing

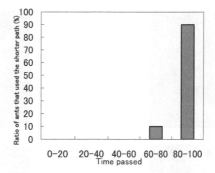

Figure 3.3: The ratio of ants that used the shorter path (adapted from Fig. 1a in [37], Springer-Verlag GmbH).

of networks. There are many unknown factors about the pheromones of actual ants; however, the volatility of pheromones can be utilized to build a model that maintains the shortest path while adapting to rapidly changing traffic. The path with a greater accumulation of pheromones is chosen at junctions, but random factors are inserted to avoid inflexible solutions in a dynamic environment.

3.1.2 Simulating the Pheromone Trails of Ants

An easy model can describe the actions of ants as follows:

- In case of nothing, random search is done.

- If the food is found, take it back to the hive. Homing ant knows the position of the hive, and returns almost straight back.

- Ants that take the food back to hive drop their pheromone. Pheromones are volatile.

- Ants not having the food have the habit of being attracted to the pheromone.

Figure 3.4 shows the simulation of the pheromone trails. Here, the hives are placed in the center, and there are three (lower right, upper left, lower left) food sources. (a) is the first random search phase. In (b), the closer lower right and lower left food is found, and the pheromone trail is formed. Upper left is in the middle of formation. In (c), pheromone trails are formed for all three sources, and makes the transport more efficient. The lower right source is almost exhaustively picked. In (d), the lower right food source finishes, and the pheromone trail has already dissipated. As a result, a vigorous transportation for the two sources on

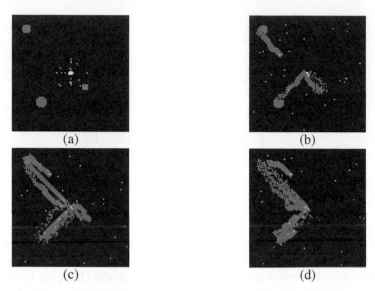

Figure 3.4: Pheromone trails of ants.

the left is being done. After this, all the sources finish, and the ants returns to random search again.

3.1.3 Generating a Death Spiral

Death spiral refers to a strange condition of the ants whereby they move endlessly akin to drawing circles, due to their attraction to pheromones released by other ants. Once they get lost without knowledge of the intended direction, they simply continue to move in circles, based on the pheromones of their company that are released from their backside. This spiral march does not stop, and they move continuously. Gradually, they weaken and die. The rest of them keep moving without taking a rest, stepping on the dead bodies, are eventually exhausted and finally die.

Although it has not been clarified why ants form the death spiral, it has been inferred that a group of lost ants may cause a death spiral. Therefore, we attempted to cause ants to form a death spiral by adding the following two phenomena to the simulation of the ant search previously described;

- During normal searching, their colony suddenly disappears, which does not allow feeding ants to return to their colony. As a result, they continue to release pheromones.

- Ants cannot grasp the location of their colony. Feeding ants show such an abnormal behavior as heading slightly off the position of the colony.

It has been reported that the death spiral can be caused by electromagnetic waves generated by smartphones, etc. The hypothesis is that some external stimulus hinders ants from grasping the correct location of their colony, resulting in them continuing to head to an incorrect position. Based on this point, the second item has been adopted. That is, we have set a position at a certain distance from the colony which lies on a line perpendicular to one connecting the current position of the ants and the actual position of the colony as an incorrect destination. This means that ants seeking their colony head for a position shifted by a certain distance from the actual position of the colony.

Figure 3.5 shows the results of the simulations described above. In (a), the figure shows the normal search of ants, and a pheromone trail is built. Next, we destroyed the colony at point (b). Consequently, many of the ants strayed, as shown in (c). After a while, what appears to be a spiral was recognized for a while (d). With the progression of time, the ants formed dumplings in (e). In (f), when they continued to chase pheromones after losing their colony, they appear to have come together in one spot instead of moving in a spiral.

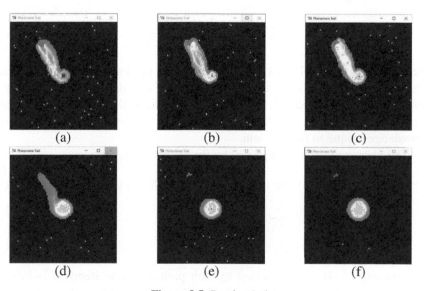

(a) (b) (c)

(d) (e) (f)

Figure 3.5: Death spirals.

3.1.4   ACO using a Pheromone Trail Model

Optimization algorithms based on the collective behavior of ants are called ant colony optimization (ACO) [30].

ACO using a pheromone trail model for the TSP uses the following algorithm to optimize the travel path:

Step1 Ants are placed randomly in each city.

Step2 Ants move to the next city. The destination is probabilistically determined based on the information on pheromones and given conditions.

Step3 Repeat until all cities are visited.

Step4 Ants that make one full cycle secrete pheromones on the route according to the length of the route.

Step5 Return to **Step1** if a satisfactory solution has not been obtained.

The ant colony optimization (ACO) algorithm can be outlined as follows. Take η_{ij} as the distance between cities i and j. The probability $p_{ij}^k(t)$ that an ant k in city i will move to city j is determined by the reciprocal of the distance $1/\eta_{ij}$ and the amount of pheromone $\tau_{ij}(t)$ as follows:

$$p_{ij}^k(t) = \frac{\tau_{ij}(t) \times \eta_{ij}^\alpha}{\sum_{h \in J_i^k} \tau_{ih}(t) \times \eta_{ih}^\alpha}. \tag{3.1}$$

Here, J_i^k is the set of all cities that the ant k in city i can move to (has not visited). The condition that ants are more likely to select a route with more pheromone reflects the positive feedback from past searches, as well as a heuristic for searching for a shorter path. The ACO can thereby include an appropriate amount of knowledge unique to the problem.

The pheromone table is updated by the following equations:

$$Q(k) \;=\; \text{the reciprocal of the path that the ant } k \text{ found} \tag{3.2}$$

$$\Delta\tau_{ij}(t) \;=\; \sum_{k \in A_{ij}} Q(k) \tag{3.3}$$

$$\tau_{ij}(t+1) \;=\; (1-\rho) \cdot \tau_{ij}(t) + \Delta\tau_{ij}(t) \tag{3.4}$$

The amount of pheromone added to each path after one iteration is inversely proportional to the length of the paths that the ants found (eq. (3.2)). The results for all ants that moved through a path are reflected in the path (eq. 3.3)). Here, A_{ij} is the set of all ants that moved on a path from city i to city j. Negative feedback to avoid local solutions is given as an evaporation coefficient (eq. (3.4)), where the amount of pheromone in the paths, or information from the past, is reduced by a fixed factor (ρ).

The ACO is an effective method for solving the traveling salesman problem (TSP) compared to other search strategies. Table 3.1 shows the optimized values for four benchmark problems and various minima found using other methods (smaller is better, obviously) [30]. The numbers in brackets indicate the number of candidates investigated. The ACO is more suitable for this problem compared to methods such as genetic algorithm (GA, section 2.1.1) simulated annealing (SA) and evolutionary programming (EP). However, it is inferior to the Lee-Kernighan method (the current TSP champion code). The characteristic that specialized methods perform better in static problems is shared by many meta-heuristics (high-level strategies which guide an underlying heuristic to increase their performance). Complicated problems, such as TSPs where the distances between cities are asymmetric or where the cities change dynamically, do not have established programs and the ACO is considered to be one of the most promising methods.

The length and pheromone accumulation of paths between cities are stored in a table (Fig. 3.6). Ants can recognize information on their surroundings, and probabilistically decide the next city to visit. The amount of pheromones added to each path after every cycle is inversely proportional to the length of the cycle path.

Table 3.1: Comparison between ACO and metaheuristics.

TSP	ACO	GA	EP	SA	Optimal
Oliver 30	420 [830]	421 [3200]	420 [40,000]	424 [24,617]	420
Eil 50	425 [1,830]	428 [25,000]	426 [100,000]	443 [68,512]	425
Eil 75	535 [3,480]	545 [80,000]	542 [325,000]	580 [173,250]	535
KroA 100	21,282 [4,820]	21,761 [103,00]	N/A [N/A]	N/A [N/A]	21,282

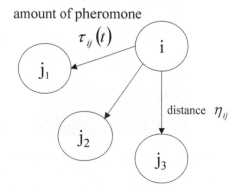

amount of pheromone

$\tau_{ij}(t)$

distance η_{ij}

Figure 3.6: Path selection rules of ants.

3.2   Particle Swarm Optimization (PSO)

3.2.1   *Collective Behavior of Boids*

Many scientists have attempted to express the group behavior of flocks of birds and schools of fish, using a variety of methods. Two of the most well-known of these scientists are Reynolds and Heppner, who simulated the movements of birds. Reynolds was fascinated by the beauty of bird flocks [98], and Heppner, a zoologist, had an interest in finding the hidden rules in the instantaneous stops and dispersions of flocks [46]. These two shared a keen understanding of the unpredictable movements of birds; at the microscopic level, the movements were extremely simple, as seen in cellular automata, while at the macroscopic level, the motions were very complicated and appeared chaotic. This is what is called an "emergent property" in the field of Artificial Life (Alife). Their model places a very large weight on the influence of individuals on each other. Similarly, it is known that an "optimal distance" is maintained among individual fish in a fish school (see Fig. 3.7).

This approach is probably not far from the mark as the basis for the social behavior of groups of birds, fish, animals and, for that matter, human beings. The sociobiologist E.O.Wilson made the interesting suggestion that the most useful information to an individual is what is shared from other members of the same group. This hypothesis forms the basis for the PSO method, which will be explained in section 3.2.2.

The collective behavior of a flock of birds emphasized the rules for keeping the optimum distance between an individual and its neighbors.

Figure 3.7: Do the movements of a school of fish follow a certain set of rules? (@Coral Sea in 2003).

The CG video by Reynolds features a group of agents called "boids." Each boid moves according to the sum of three vectors: (1) force to move away from the nearest individual or obstacle, (2) force to move toward the center of the flock, and (3) force to move toward its destination. Adjusting coefficients in this summation results in many behavioral patterns. This technique is often used in special effects and videos in films. Figure 3.8 (simple behavior of flocks) and Fig. 3.9 (situation with obstacles) are examples of simulations of boids.

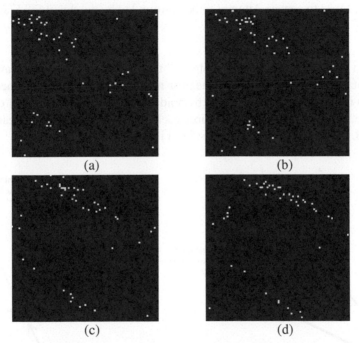

(a)   (b)

(c)   (d)

Figure 3.8: Simple behavior of boids ((a)⇒(b)⇒(c)⇒(d)).

(a)   (b)

Figure 3.9: Boids in a situation with obstacles ((a)⇒(b)).

The following are the details of the algorithm that boids follow. Many individuals (boids) move around in space, and each individual has a velocity vector. The three factors below result in a flock of boids.

1. Avoid collision: Attempt to avoid collision with nearby individuals.

2. Match pace: Attempt to match velocity of nearby individuals.

3. Move to center: Attempt to be surrounded by nearby individuals.

Each boid has an "optimum distance" to avoid collision, and behaves so as to maintain this distance with its nearest neighbor [102]. Collision becomes a concern if the distance between nearby boids becomes shorter than the "optimum distance." Therefore, to avoid collision, each boid slows down if the nearest boid is ahead, and speeds up if the nearest boid is behind (Fig. 3.10).

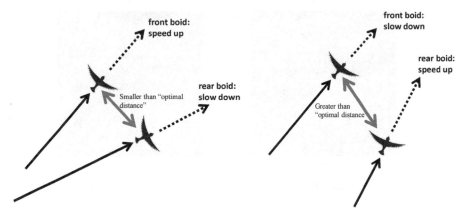

Figure 3.10: Avoid collision (1).　　　　**Figure 3.11:** Avoid collision (2).

The "optimum distance" is also used to prevent the risk of straying from the flock. If the distance to the nearest boid is larger than the "optimum distance," each boid speeds up if the nearest boid is ahead, and slows down if it is behind (Fig. 3.11).

Here, "ahead" and "behind" are defined as ahead or behind a line that crosses the boid's eyes and is perpendicular to the direction in which the boid is moving (Fig. 3.12). Boids try to move parallel to (with the same vector as) their nearest neighbor. Here, there is no change in speed. Furthermore, boids change velocity so as to always move toward the center of the flock (center of gravity of all boids).

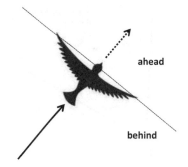

Figure 3.12: Ahead or behind a line that crosses the boid's eyes.

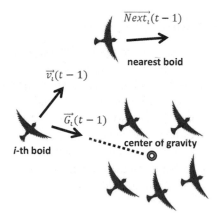

Figure 3.13: Updating the velocity vector.

In summary, the velocity vector ($\vec{v}_i(t)$) of the i-th boid is updated at time t as follows (see Fig. 3.13):

$$\vec{v}_i(t) = \vec{v}_i(t-1) + \vec{Next}_i(t-1) + \vec{G}_i(t-1). \tag{3.5}$$

Where $\vec{Next}_i(t-1)$ is the velocity vector of the nearest boid to individual i, and $\vec{G}_i(t-1)$ is the vector from individual i to the center of gravity. The velocity at one step before, i.e., $\vec{v}_i(t-1)$, is added to take inertia into account.

Each boid has its own field of view (Fig. 3.14) and considers boids within its view when finding the nearest neighbor. However, the coordinates of all boids, including those out of view, are used to calculate the center of gravity.

Kennedy et al. designed an effective optimization algorithm using the mechanism behind boids [62]. This is called particle swarm optimization (PSO), and numerous applications are reported. The details are provided in section 3.2.2.

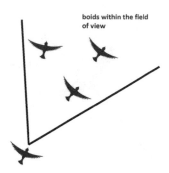

boids within the field of view

Figure 3.14: Each boid has its own field of view.

3.2.2 PSO Algorithm

The classic PSO was intended to be applied to optimization problems. It simulates the motion of a large number of individuals (or "particles") moving in a multi-dimensional space [62]. Each individual stores its own location vector ($\vec{x}_i$), velocity vector ($\vec{v}_i$), and the position at which the individual obtained the highest fitness value ($\vec{p}_i$). All individuals also share information regarding the position with the highest fitness value for the group ($\vec{p}_g$).

As generations progress, the velocity of each individual is updated using the best overall location obtained up to the current time for the entire group and the best locations obtained up to the current time for that individual. This update is performed using the following formula:

$$\vec{v}_i = \chi(\omega\vec{v}_i + \phi_1 \cdot (\vec{p}_i - \vec{x}_i) + \phi_2 \cdot (\vec{p}_g - \vec{x}_i)) \tag{3.6}$$

The coefficients employed here are the convergence coefficient χ (a random value between 0.9 and 1.0) and the attenuation coefficient ω, while ϕ_1 and ϕ_2 are random values unique to each individual and the dimension, with a maximum value of 2. When the calculated velocity exceeds some limit, it is replaced by a maximum velocity V_{max}. This procedure allows us to hold the individuals within the search region during the search.

The locations of each of the individuals are updated at each generation by the following formula:

$$\vec{x}_i = \vec{x}_i + \vec{v}_i \tag{3.7}$$

The overall flow of the PSO is as shown in Fig. 3.15. Let us now consider the specific movements of each individual (see Fig. 3.16). A flock consisting of a number of birds is assumed to be in flight. We focus on one of the individuals (Step 1). In the figure, the $\bigcirc$ symbols and linking line segments indicate the positions and paths of the bird. The nearby $\odot$ symbol (on its path) indicates the position with the highest fitness value on the individual's path (Step 2). The distant $\odot$ symbol (on the other bird's path) marks the position with the highest

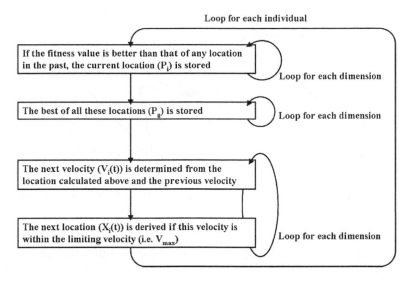

Figure 3.15: Flow chart of the PSO algorithm.

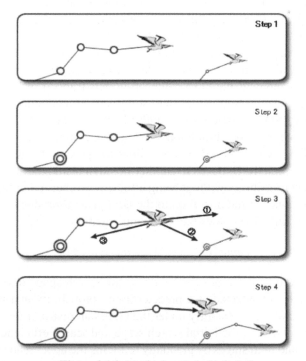

Figure 3.16: In which way do birds fly?

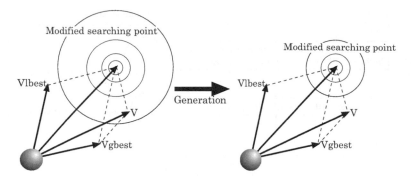

Figure 3.17: Concept of searching process by PSO with Gaussian mutation.

fitness value for the flock (Step 2). One would expect that the next state will be reached in the direction shown by the arrows in Step 3. Vector ① shows the direction followed in the previous steps; vector ② is directed towards the position with the highest fitness for the flock; and vector ③ points to the location where the individual obtained its highest fitness value so far. Thus, all these vectors, ①, ② and ③, in Step 3 are summed in order to obtain the actual direction of movement in the subsequent step (see Step 4).

The efficiency of this type of PSO search is certainly high because focused searching is available near optimal solutions in a relatively simple search space. However, the canonical PSO algorithm often gets trapped in local optimum in multimodal problems. Because of that, some sort of adaptation is necessary in order to apply PSO to problems with multiple sharp peaks.

To overcome the above limitation, a GA-like mutation can be integrated with PSO [48]. This hybrid PSO does not follow the process by which every individual of the simple PSO moves to another position inside the search area with a predetermined probability without being affected by other individuals, but leaves a certain ambiguity in the transition to the next generation due to Gaussian mutation. This technique employs the following equation:

$$mut(x) = x \times (1 + gaussian(\sigma)), \tag{3.8}$$

where σ is set to be 0.1 times the length of the search space in one dimension. The individuals are selected at a predetermined probability and their positions are determined at the probability under the Gaussian distribution. Wide-ranging searches are possible at the initial search stage and search efficiency is improved at the middle and final stages by gradually reducing the appearance ratio of Gaussian mutation at the initial stage. Figure 3.17 shows the PSO search process with Gaussian mutation. In the figure, V_{lbest} represents the velocity based on the local best, i.e. $\vec{p}_i - \vec{x}_i$ in Eq. (3.6), whereas V_{gbest} represents the velocity based on the global best, i.e. $\vec{p}_g - \vec{x}_i$.

3.2.3 Comparison with GA

Let us turn to a comparison of the performance of the PSO with that of the GA using benchmark functions, in order to examine the effectiveness of PSO.

For the comparison, $F8$ (Rastrigin's function) and $F9$ (Griewangk's function) are employed. These are defined as:

$$F8(x_1,x_2) \;=\; 20 + x_1^2 - 10\cos(2\pi x_1) + x_2^2 - 10\cos(2\pi x_2) - (-5.11 \le x_i \le 5.11)$$

$$F9(x_1,x_2) \;=\; \frac{1}{4000}\sum_{i=1}^{2}(x_i - 100)^2 - \prod_{i=1}^{2}\cos\left(\frac{x_i - 100}{\sqrt{i}}\right) + 1(-10 \le x_i \le 10)$$

Figure 3.18 and Fig. 3.19 show the shapes of $F8$ and $F9$, respectively. $F8$ and $F9$ seek the minimum value. $F8$ contains a large number of peaks so its optimization is particularly difficult.

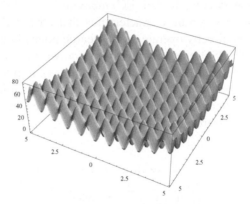

Figure 3.18: Rastrigin's function ($F8$).

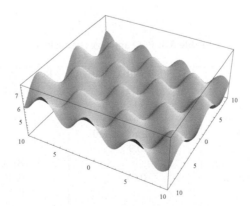

Figure 3.19: Griewangk's function ($F9$).

Comparative experiments were conducted with PSO and GA using the above benchmark functions. PSO and GA were repeatedly run 100 times. Search space ranges for the experiments are listed in Table 3.2. PSO and GA parameters are given in Table 3.3.

The performance results are shown in Figs. 3.20 and 3.21 which plot the fitness values against the generations. Table 3.4 shows the averaged best fitness values over 100 runs. As can be seen from the table and the figures, the combination of PSO with Gaussian mutation allows us to achieve a performance that is almost equal to that of the canonical PSO for the unimodals, and a better performance than the canonical PSO for the multimodals. The experimental results with other benchmark functions are further discussed in [53].

PSO is a stochastic search method, as are GA and GP, and its method of adjustment of $\vec{p}_i$ and $\vec{p}_g$ resembles crossover in GA. It also employs the concept of fitness, as in evolutionary computation. Thus, the PSO algorithm is strongly related to evolutionary computation (EC) methods.

However, PSO has certain characteristics that other EC techniques do not have. GA operators directly operate on the search points in a multi-dimensional search space, while PSO operates on the motion vectors of particles which, in turn, update the search points (i.e., particle positions). In other words, GA operators are position specific and the PSO operators are direction specific. One of the reasons PSO gathered so much attention was the tendency of its individuals to proceed directly towards the target.

Table 3.2: Search space for test functions.

Function	Search space
F8	$-65.535 \leq x_i < 65.536$
F9	$-10 \leq x_i \leq 10$

Table 3.3: PSO and GA parameters.

Parameters	PSO,PSO with Gaussian	Real-valued GA
Population	200	200
V_{max}	1	
Generation	50	50
ϕ_1,ϕ_2	upper limits = 2.0	
Inertia weight	0.9	
Crossover ratio		0.7(BLX-α)
Mutation	0.01	0.01
Elite		0.05
Selection		tournament (size=6)

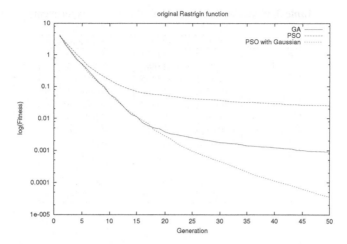

Figure 3.20: Standard PSO versus PSO with Gaussian mutation for $F8$.

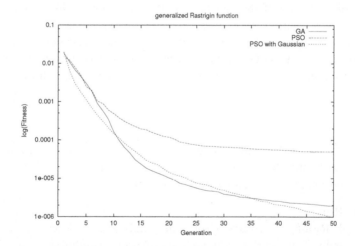

Figure 3.21: Standard PSO versus PSO with Gaussian mutation for $F9$.

In the chapter "The Optimal Allocation of Trials" in his book [51], Holland ascribes the success of EC on the balance of "exploitation," through search of known regions, with "exploration," through search, at finite risks, of unknown regions. PSO is adept at managing such subtle balances. These stochastic factors enable PSO to make thorough searches of the relatively promising regions and, due to the momentum of speed, also allows effective searches of unknown regions. Theoretical research is currently underway to derive optimized values for PSO parameters by mathematical analysis, for stability and convergence (see [18, 63]).

Table 3.4: Average best fitness of 100 runs for experiments.

	Gen	GA	PSO	PSO with gaussian
$F8$	1	4.290568	3.936564	3.913959
	10	0.05674	0.16096	0.057193
	20	0.003755	0.052005	0.002797
	30	0.001759	0.037106	0.000454
	40	0.001226	0.029099	0.000113
	50	0.000916	0.02492	3.61E-05
$F9$	1	0.018524	0.015017	0.019726
	10	0.000161	0.000484	0.000145
	20	1.02E-05	0.000118	1.43E-05
	30	3.87E-06	6.54E-05	4.92E-06
	40	2.55E-06	5.50E-05	2.04E-06
	50	1.93E-06	4.95E-05	1.00E-06

3.2.4 Collective Memory and Spatial Sorting

Boid algorithm is simple and the emergence of group actions can be easily realized, however, one of the problems is that it is impossible to control group behavior. In this section, we consider creating more realistic Boid.

For this purpose, the concept of collective memory has been introduced [21]. This concept considers that the past records of group structure affect interactions among individuals. The basic principles of this model are as follows:

■ **Rule1**
All individuals always maintain a minimum distance from each other (avoidance behavior). This action has the highest priority and is also observed in the actions of real animals.

■ **Rule2**
If **Rule1** is not carried out, individuals tend to be attracted to other individuals and lined up with neighboring individuals. It is related to isolation avoidance.

Each boid's surroundings can be modeled as shown in Fig. 3.22. Here, the target individual is drawn in the shape of the bird, and its surroundings are divided from the closest one as follows:

■ Repulsion: Two individuals in this zone repel each other (abbr. Zor = zone of repulsion)

■ Orientation: Two individuals in this zone are lined up in the same direction (abbr. Zoo = zone of orientation)

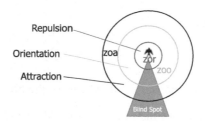

Figure 3.22: Modeling of surroundings.

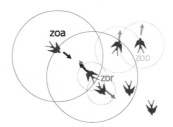

Figure 3.23: Local interaction.

■ Attraction: Two individuals in this zone are attracted to each other (abbr. Zoa = zone of attraction)

There is a blind spot behind oneself. Interactions between individuals are shown in Fig. 3.23.

To realize the above **Rule1** and **Rule2** with local rules, the update rule is set as follows. The update is performed for each individual i. The position vector of individual i is $\vec{c}_i$ and the velocity vector is $\vec{v}_i$.

■ When there are individuals in Zor (zone of repulsion):
The next direction to move is,

$$\vec{d}_r(t+\tau) = -\sum_{j \neq i}^{n_r} \frac{\vec{r}_{ij}(t)}{|\vec{r}_{ij}(t)|}. \tag{3.9}$$

Vector $\vec{r}_{ij} = \frac{\vec{c}_j - \vec{c}_i}{|\vec{c}_j - \vec{c}_i|}$ is a unit vector in the direction of an individual j. This is of the highest priority. Thus, $\vec{d}_i(t+\tau) = \vec{d}_r(t+\tau)$.

■ When there are no individuals in Zor (zone of repulsion):
First derive the following vectors:

■ Against Zoo:

$$\vec{d}_o(t+\tau) = \sum_{j=1}^{n_o} \frac{\vec{v}_j(t)}{|\vec{v}_j(t)|}, \tag{3.10}$$

■ Against Zoa:

$$\vec{d}_a(t + \tau) = \sum_{j \neq i}^{n_a} \frac{\vec{r}_{ij}(t)}{|\vec{r}_{ij}(t)|}. \tag{3.11}$$

The vectors are summed except for the blind spot.

Then the direction of the movement is calculated in the following way:

■ When individuals exist only in Zoo:

$$\vec{d}_i(t + \tau) = \vec{d}_o(t + \tau) \tag{3.12}$$

■ When individuals exist only in Zoa:

$$\vec{d}_i(t + \tau) = \vec{d}_a(t + \tau) \tag{3.13}$$

■ When individuals exist in both:

$$\vec{d}_i(t + \tau) = \frac{1}{2}(\vec{d}_o(t + \tau) + \vec{d}_a(t + \tau)) \tag{3.14}$$

■ When there are no individuals in any of them:

$$\vec{d}_i(t + \tau) = \vec{v}_i(t) \tag{3.15}$$

Note that the Zoo (zone of orientation) changes with time.
Finally, the following adjustments are made.

■ Move $\vec{d}_i(t + \tau)$ by a random angle based on the Gaussian distribution.

■ Each individual i is redirected by a vector $\vec{v}_i(t + \tau) = \vec{d}_i(t + \tau)$. However, when the maximum rotation angle θ_τ is exceeded, it is set to be this maximum angle.

■ The movement speed of each individual is set to be constant s.

The parameters of the simulations are presented in Table 3.5. Units represent the scale of the length of the specific organism of interest. For example, it is very small for insects, and other parameters are scaled appropriately.

Figure 3.24 shows the experimental results under various conditions. The chosen parameter values are: $N = 100, r_r = 1, \alpha = 270, \theta = 40, s = 3$ and $\sigma = 0.05$. Collective behavior is classified in the following four cases:

■ (A) swarming

■ (B) torus

Table 3.5: Summary of model parameters.

Parameter	Unit	Symbol	Values explored
Number of individuals	None	N	10–100
Zone of repulsion	Units	r_r	1
Zone of orientation	Units	$\Delta r_o(r_o - r_r)$	0–15
Zone of attraction	Units	$\Delta r_a(r_a - r_o)$	0–15
Field of perception	Degrees	α	200–360
Turning rate	Degrees per second	θ	10–100
Speed	Units per second	s	1–5
Error (std. deviation)	Degrees (rad)	σ	0–11.5(0–0.2 rad)
Time step increment	Seconds	τ	0.1

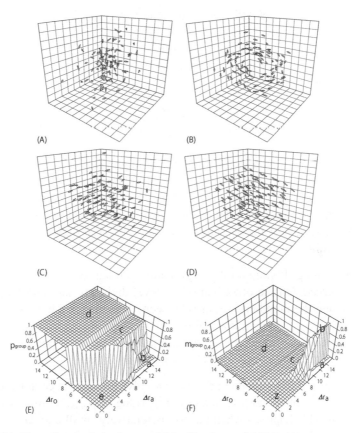

Figure 3.24: The collective behaviours exhibited by the model [21]: (A) swarming, (B) torus, (C) dynamic parallel group, (D) highly parallel group. Also shown are the group polarization p_{group} (E) and angular momentum m_{group} (F) as a function of changes in the size of the zone of orientation Δr_o and zone of attraction Δr_a.

- ■ (C) dynamic parallel group

- ■ (D) highly parallel group

The lower part shows values (average of 30 runs) of the group polarization p_{group} (E) and angular momentum m_{group} (F) as a function of the change in the size of the zone of orientation and zone of attraction (Δr_o and Δr_a). Parameter domain, presented in the figure from a to d, corresponds to collective behavior presented above. Region e is the division action, occurring with at least 50% probability.

The group polarization p_{group} (E) and angular momentum m_{group} (F) are the global properties of the model and are defined as follows:

$$p_{group}(t) = \frac{1}{N} \left| \sum_{i=1}^{N} \vec{v}_i(t) \right| \tag{3.16}$$

$$m_{group}(t) = \frac{1}{N} \left| \sum_{i=1}^{N} \vec{r}_{ic}(t) \times \vec{v}_i(t) \right| \tag{3.17}$$

$$\vec{r}_{ic} = \vec{c}_i - \vec{c}_{group} \tag{3.18}$$

$$\vec{c}_{group}(t) = \frac{1}{N} \sum_{i=1}^{N} \vec{c}_i(t) \tag{3.19}$$

Group polarization increases as the degree of alignment between individuals in a population is increased. On the other hand, angular momentum is a measurement of the degree of rotation of the center of gravity of the population and is the sum of the angular momenta of individuals around the center of gravity.

From the results of these figures, it can be observed that the behavior of the group is as follows:

- ■ Swarming: Population by aggregation (an aggregate with cohesion) are observed. On the other hand, low-level divisions and parallel alignment among the members can also be seen. This means a low degree of solidarity (p_{group}) and a low angular momentum (m_{group}). This indicates that the individuals take the action of repulsion and attraction, but they occur when they are not parallel (Region a of the figure).

- ■ Torus: An individual continues to revolve around the center forever. The direction of rotation is random. The value of p_{group} is low, but m_{group} is high. This happens when the Δr_o is relatively small and Δr_a is relatively large (Region b).

- ■ Dynamic parallel group: This group shows high p_{group} and low m_{group}. This kind of population is more mobile than swarming or torus. This phenomenon occurs for medium Δr_o and a medium or higher Δr_a (Region c).

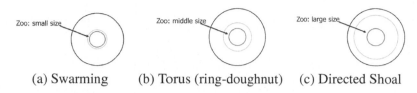

(a) Swarming (b) Torus (ring-doughnut) (c) Directed Shoal

Figure 3.25: Different sizes of Zoo.

■ Highly parallel group: When Δr_o increases, the population undergoes self-organization and transitions into highly aligned state (low m_{group}) with linear motion (low m_{group}) (Region d).

We present the results of a simulation using the above method. This simulation is created based on Boid. As mentioned earlier, each Boid has three radii and one angle. Those are radius R_{Zor} representing the range of the Repulsion, R_{Zoo} representing the range of the Orientation, R_{Zoa} representing the range of the attraction and viewing angle (field of perception) α_p. Radii are defined to satisfy the following inequalities:

$$0 \le R_{Zor} \le R_{Zoo} \le R_{Zoa}. \tag{3.20}$$

In the region where R_{Zoo} is small, the swarming state in which Boids stay in place, without moving in parallel or rotating in the same direction, was observed (Fig. 3.25(a)). This is shown in Fig. 3.26(a). It can be seen that three swarms are created in the upper left, lower left and lower right. Those groups are closer to each other because the edges of the screen are connected, however, swarms occasionally exchanged Boids without being combined.

In the intermediate region of R_{Zoo} (i.e., where R_{Zoo} is of middle size), it was observed that Boids circled around the empty core in the same direction (see Fig. 3.25(b) and Fig. 3.26(b)). In the figure, a large torus is formed on the left side. Torus is unstable, and often takes the form of swarming or parallel group after some time.

When R_{Zoo} is relatively big, the Boids have been observed to form a group and move in parallel (see Fig. 3.25(c) and Fig. 3.26(c)). On the right side of the figure, it can be seen that the Boids became a group and move in parallel. If R_{Zoo} is small, the resulting parallel group quickly self-destructs, becoming a swarm, and repeats to become a parallel group again. On the other hand, if R_{Zoo} is large, the formed parallel group continues to move in a certain direction without self-destructing.

3.2.5 Boids Attacked by an Enemies

Let us realize the escape behavior or evasive maneuver in Boids. This is equivalent to the behavior of small fish running away from a big enemy. In the natural

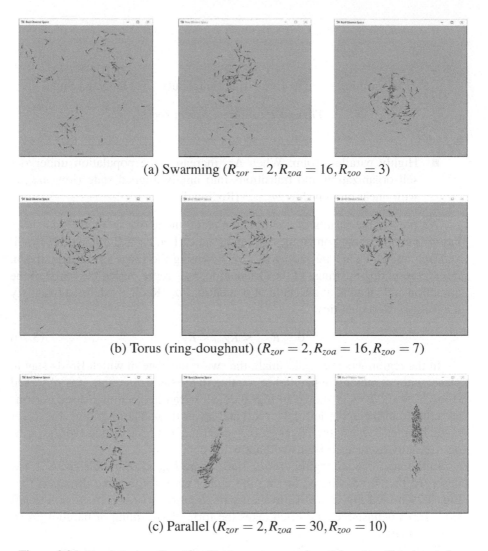

(a) Swarming ($R_{zor} = 2, R_{zoa} = 16, R_{zoo} = 3$)

(b) Torus (ring-doughnut) ($R_{zor} = 2, R_{zoa} = 16, R_{zoo} = 7$)

(c) Parallel ($R_{zor} = 2, R_{zoa} = 30, R_{zoo} = 10$)

Figure 3.26: Simulation results with collective memory and spatial sorting. Time lapses from left to right.

world, it can be seen that a school panics and disperses. For example, the following fish behaviors are well-known (see Fig. 3.27):

■ Dolphins attack bait ball in South Africa

■ Sardine run

■ Rainbow runner attacks a school of sardines

■ Mackerel tuna attacks a school of sardines

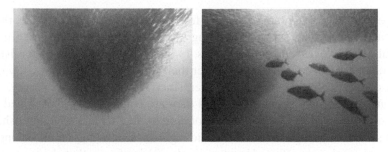

Figure 3.27: School of sardines attacked by GT (@Moal Boal, Philippines in 2016).

Among the behaviors observed in those cases, the main ones are:

■ Flash expansion (Fig. 3.28(a))

■ Fountain effect (Fig. 3.28(b))

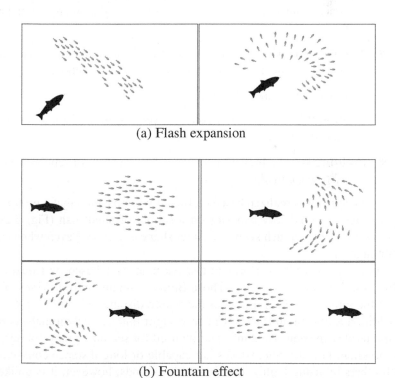

(a) Flash expansion

(b) Fountain effect

Figure 3.28: How to get away from a large predator.

Large fish are slow in movement and few in numbers. To compensate, they can see far away and have a habit of going towards a school of small fish in sight. Small fish move fast and are numerous. In order to escape from bigger fish, their natural enemies, the moment they enter their field of vision, they swiftly change their course and try to escape. However, since they can see only nearby, the appearance of big fish is unnoticed until they are considerably approached (those are called "instincts"). Individuals belonging to the same species (big fish, small fish) swim according to the Boids basic algorithm, and individuals belonging to different species are affected by attractive/repulsive forces that change according to the "instincts" mentioned above. In order to realize this, the Boids algorithm was revised and the rules of interaction were added to the three basic rules for the same kind. The three basic rules are as follows:

- **Heading to the center** Easier to point in the direction, where more individuals belonging to the same group reside.

- **Aligning position** Approach to be in the same position as the fish in the same group. Basically, it works as an attraction. When it gets too close, it becomes repulsion, so that they do not perfectly group up around a single point.

- **Aligning direction** A force of attraction works in an angular space that faces in the same direction as the fish of the same school.

The rules of interaction are as follows:

- **hunting** When a smaller opponent is seen, a force of attraction is experienced (hunt mode).

- **avoiding** When a bigger opponent is seen, a force of repulsion is experienced (escape mode).

When Boids was realized based on this, it was observed that a group of large fish engaged in predatory behavior on a school of small fish (Fig. 3.29). It is noteworthy that small fish keep a constant shape as a school even when escaping from larger fish.

Let us create Boids performing evasive maneuver based on Couzin's algorithm described in section 3.2.4. Those Boids simulate multiple kinds of behavior at the same time. Each Boid has the search distance R' (fulfilling $R_{Zor} \geq R'$) in addition to the parameters of Couzin's algorithm. In real animals, communication used with friends and information used for searching may not necessarily be the same. For example, whales are capable of long-distance communication with others by using highly directional ultrasounds, however, it is unlikely that echolocation will be able to reach that far. Therefore, the R' is smaller than R_{Zor}.

If there are two individuals of different species, they are seen as a predator and non-predator (prey). In this case, the update rules are set as follows. This update is performed for each individual i.

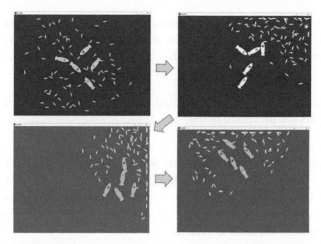

Figure 3.29: Simulation results. A school of small fish attacked by a few large fish.

1. When an individual of different species is in Zor (zone of repulsion), the direction to move next is as follows:

$$\vec{d_r}(t + \tau) = \perp \sum_{j \neq i, sp(i) \neq sp(j)}^{n_r} \frac{\vec{r}_{ij}(t)}{|\vec{r}_{ij}(t)|}. \qquad (3.21)$$

$sp(i)$ means species of the individual i. The right side is the direction to the center of gravity of heterogeneous individuals (individuals from different species). Here, $\pm$ is $-$ for prey species and $+$ for predators. Thus, it is possible to simulate the situation of "the chasing side and the escaping side."

2. When an individual of different species is in Zor (zone of repulsion), i.e., when there are heterogeneous individuals within the enemy search range (a circle with a radius of R' that is limited by the viewing angle), the direction to move next is as follows:

$$\vec{d_r}(t + \tau) = \pm \sum_{j \neq i, sp(i) \neq sp(j)}^{N'} \frac{\vec{r}_{ij}(t)}{|\vec{r}_{ij}(t)|}. \qquad (3.22)$$

This is the same derivation as the above, except that N' is the number of individuals in the enemy search range.

3. When it is neither of the above, Couzin's algorithm is executed. Observations at this time are done only for individuals of the same species.

4. If a heterogeneous individual is found through the above processes, the speed of the next step is multiplied by α, otherwise by α^{-1}. Maximum

rotation angle θ_τ is set to different values for the case when a heterogeneous individual is found and for the case when it is not. This is to realize that there is a difference in behavior when escaping from an enemy and during normal times.

By simulating the enemy approaching, the Boid evasive maneuvers were simulated. Results are presented below.

■ Flash expansion
It was observed that the Boids acting in parallel spread out to escape from the enemies. Figure 3.30(a) shows the situation when the tail was attacked. Red Boid is a predator species and blue Boid is a non-predator (prey) species. While evasive maneuver is being carried out, the prey species are depicted in green.

■ Fountain effect
Figure 3.30(b) depicts a situation when prey is attacked from the side. As can be seen from the figure, it seems that the Boids spread out when attacked and try to minimize the sacrifices.

3.3   Artificial Bee Colony Optimization (ABC)

Bees, along with ants, are well-known examples of social insects (Fig. 3.31). Bees are classified into three types:

■ employed bees

■ onlooker bees

■ and scout bees

Employed bees fly in the vicinity of feeding sites they have identified, sending information about food to onlooker bees. Onlooker bees use the information from employed bees to perform selective searches for the best food sources from the feeding site. When information about a feeding site is not updated for a given period of time, its employed bees abandon it and become scout bees that search for a new feeding site. The objective of a bee colony is to find the highest-rated feeding sites. The population is approximately half employed bees and scout bees (about 10–15% of the total), the rest are onlooker bees.

The waggle dance (a series of movements) performed by employed bees to transmit information to onlooker bees is well known (Fig. 3.32). The dance involves shaking the hindquarters and indicating the angle with which the sun will

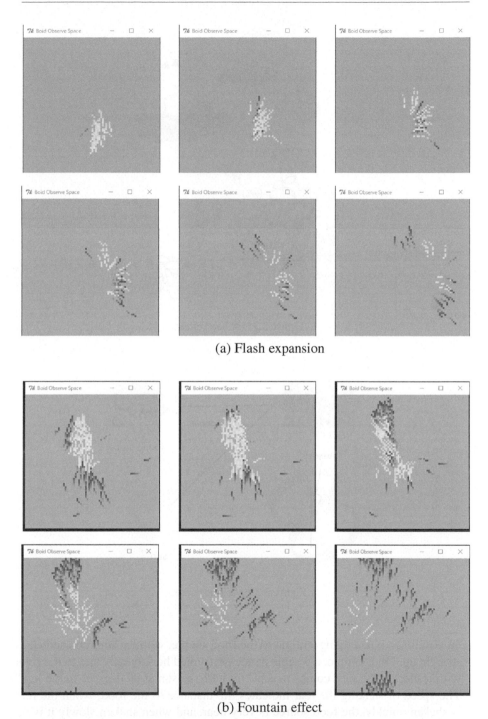

(a) Flash expansion

(b) Fountain effect

Figure 3.30: Evasive maneuver simulation. Time lapses from top left to bottom right.

Figure 3.31: A bee colony.

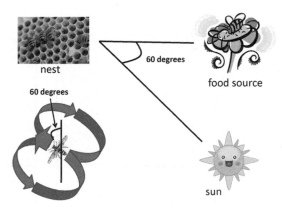

Figure 3.32: Waggle dance.

be positioned when flying straight to the food source, with the sun represented as straight up. For example, a waggle dance performed horizontally and to the right with respect to the nest combs means "fly with the sun at 90 degrees to the left." The speed of shaking the rear indicates the distance to the food; when the rear is shaken quickly, the food source is very near, and when shaken slowly it is far away. Communication via similar dances is also performed with regard to pollen and water collection, as well as the selection of locations for new hives.

The Artificial Bee Colony (ABC) algorithm [58, 59], initially proposed by Karaboga et al., is a swarm optimization algorithm that mimics the foraging behavior of honey bees. Since ABC was designed, it has been proved that ABC, with fewer control parameters, is very effective and competitive with other search techniques, such as genetic algorithm (GA), particle swarm optimization (PSO), and differential evolution (DE).

In ABC algorithms, an artificial swarm is divided into employed bees, onlooker bees, and scouts. N d-dimensional solution candidates to the problem are randomly initialized in the domain and referred to as food sources. Each employed bee is assigned to a specific food source $\vec{x}_i$ and searches for a new food source $\vec{v}_i$ by using the following operator:

$$\vec{v}_{ij} = \vec{x}_{ij} + \text{rand}(-1, 1) \times (\vec{x}_{ij} - \vec{x}_{kj}), \tag{3.23}$$

where $k \in \{1, 2, \cdots, N\}, k \neq i$, and $j \in \{1, 2, \cdots, d\}$ are randomly chosen indices. $\vec{v}_{ij}$ is the jth element of the vector $\vec{v}_i$. If the trail to a food source is outside of the domain, it is reset to an acceptable value. The obtained $\vec{v}_i$ is then evaluated and put into competition with $\vec{x}_i$ for survival. The bee prefers the better food source. Unlike employed bees, each onlooker bee chooses a preferable source according to the food source's fitness in order to do further searches in the food space using eq. (3.23). This preference scheme is based on the fitness feedback information from employed bees. In classic ABC [58], the probability that the food source $\vec{x}_i$ can be exploited is expressed as

$$p_i = \frac{fit_i}{\sum_{j=1}^{N} fit_j}, \tag{3.24}$$

where fit_i is the fitness of the ith food source, $\vec{x}_i$. For the sake of simplicity, we assume that the fitness value is non-negative and that the larger, the better. If the trail $\vec{v}_i$ is superior to $\vec{x}_i$ in terms of profitability, this onlooker bee informs the relevant employed bee associated with the ith food source, $\vec{x}_i$, to renew its memory and forget the old one. If a food source cannot be improved upon within a predetermined number of iterations, defined as Limit, this food source is abandoned. The bee that was exploiting this food site becomes a scout and associates itself with a new food site that is chosen via some principle. In canonical ABC [58], the scout looks for a new food site by random initialization.

The details of the ABC algorithm is described as below. The pseudocode of the algorithm is shown in Algorithm 1.

Step 0: Preparation The total number of search points (N), total number of trips (T_{max}), and a scout control parameter (Limit) are initialized. The numbers of employed bees and onlooker bees are set to be the same as the total number of search points (N). The value of the objective function f is taken to be nonnegative, with larger values being better.

Algorithm 1 The ABC algorithm

Require: T_{max}, #. of employed bees ($=$ #. of onlooker bees), `Limit`
   Initialize food sources
   Evaluate food sources
   $i = 1$
  while $i < T_{max}$ **do**
     Use employed bees to produce new solutions
     Evaluate the new solutions and apply greedy selection process
     Calculate the probability values using fitness values
     Use onlooker bees to produce new solutions
     Evaluate new solutions and apply greedy selection process
     Determine abandoned solutions and use scouts to generate new ones randomly
     Remember the best solution found so far
     $i = i + 1$
  end while
   Return best solution

Step 1: Initialization 1 The trip counter k is set to 1, and the number of search point updates s_i is set to 0. The initial position vector for each search point $\vec{x}_i = (x_{i1}, x_{i2}, x_{i3}, \cdots, x_{id})^T$ is assigned random values. Here, the subscript i $(i = 1, \cdots, N)$ is the index of the search point, and d is the number of dimensions in the search space.

Step 2: Initialization 2 Determine the initial best solution $\vec{best}$.

$$i_g = \operatorname*{argmax}_{i} f(\vec{x}_i) \qquad (3.25)$$

$$\vec{best} = x_{i_g} \qquad (3.26)$$

Step 3: Employed bee search The following equation is used to calculate a new position vector $\vec{v}_{ij}$ from the current position vector $\vec{x}_{ij}$.

$$\vec{v}_{ij} = \vec{x}_{ij} + \phi \cdot (\vec{x}_{ij} - \vec{x}_{kj}), \qquad (3.27)$$

Here, j is a randomly chosen dimensional number, k is the index for some randomly chosen search point other than i, and ϕ is a uniform random number in the range $[-1, 1]$. The position vector $\vec{x}_i$ and the number of search point updates s_i are determined according to the following equation:

$$I = \{i \mid f(\vec{x}_i) < f(\vec{v}_i)\} \qquad (3.28)$$

$$\vec{x}_i = \begin{cases} \vec{v}_i & i \in I \\ \vec{x}_i & i \notin I \end{cases} \qquad (3.29)$$

$$s_i = \begin{cases} 0 & i \in I \\ s_i + 1 & i \notin I \end{cases} \qquad (3.30)$$

Step 4: Onlooker bee search The following two steps are performed.

1. Relative ranking of search points
   The relative probability P_i is calculated from the fitness fit_i, which is based on the evaluation score of each search point. Note that $fit_i = f(\vec{x}_i)$. The onlooker bee search counter l is set to 1.

$$P_i = \frac{fit_i}{\sum_{j=1}^{N} fit_j} \qquad (3.31)$$

2. Roulette selection and search point updating
   Search points are selected for updating based on the probability P_i, calculated above. After search points have been selected, perform a procedure as in **Step 3** to update the search point position vectors. Then, let $l = l + 1$ and repeat until $l = N$.

Step 5: Scout bee search Given a search point for which $s_i \geq \texttt{Limit}$, random numbers are used to exchange generated search points.

Step 6: Update best solution Update the best solution $\vec{best}$.

$$i_g = \underset{i}{\mathrm{argmax}}\, f(\vec{x}_i) \qquad (3.32)$$

$$\vec{best} = \vec{x}_{i_g} \text{ when } f(x_{i_g}) > f(\vec{best}) \qquad (3.33)$$

$$(3.34)$$

Step 7: End determination End if $k = T_{max}$. Otherwise, let $k = k + 1$ and return to **Step 3**.

The ABC bee search process is explained by a simple example. Figure 3.33 depicts a bee colony and the food ground, which is the search space. Four employed bees, four onlooker bees, and one scout bee are in the colony. Suppose that good search segments (higher degree of fitness) are those shown in dark color. In Fig. 3.33, there are two hills, with the one on the bottom-right representing the optimum value. Suppose that food grounds 1 to 4 are assigned to each employed bee upon initialization.

Each employed bee updates its assigned search position (food ground) once (an action corresponding to an employed bee visiting the food ground once and performing a quick exploration of the surroundings). In Fig. 3.34, bees 2 and 3 out of the 4 employed bees have updated their positions and moved to a food ground of higher fitness. However, this change is probabilistic and, therefore, does not necessarily take them to a place of higher fitness. For this reason, employed bees 1 and 4 remain in the same place. In other words, they are not performing a local search.

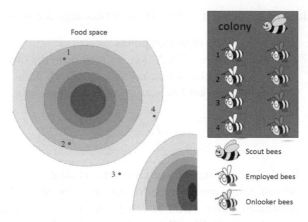

Figure 3.33: Bee colonies and search space.

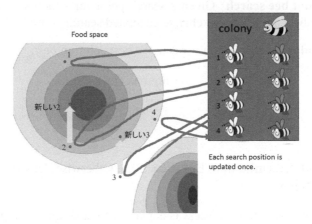

Figure 3.34: Employed bees search.

Next, the onlooker bee performs a search. In this case, food grounds with higher degrees of fitness are more likely to be visited, and, thus, more easily selected during an update. In Fig. 3.35, onlooker bees 1 and 2 update search position 2 (the food ground of employed bee 2) and 3 (the food ground for employed bee 3), respectively. Since positions with higher degrees of fitness are preferred, onlooker bees 3 and 4 also visit search position 2, but no update occurs because it is already a local maximum (Fig. 3.36).

After that, it is the scout bees' turn to perform the search. This bee resets food grounds that have not been updated for a fixed number of times. In this case, search position 2 is selected and replaced by a position displaced at random (Fig. 3.37). Obviously, the food ground for employed bee 2 is also updated to this randomly displaced position.

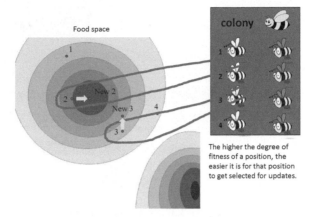

Figure 3.35: Onlooker bees' search (1).

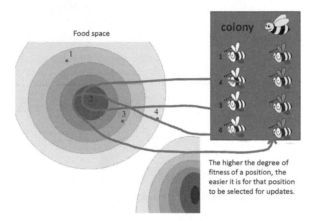

Figure 3.36: Onlooker bees' search (2).

This completes a cycle of actions in the colony. Subsequently, employed bees start the search again. In Fig. 3.38, 3 employed bees update their food grounds. Onlooker bees then search preferentially for food grounds with a higher degree of fitness. As a result, food ground 2 becomes the optimal value (Fig. 3.39).

ABC has recently been improved in many aspects. For instance, we analyzed the mechanism of ABC to show a possible drawback of using parameter perturbation. To overcome this deficiency, we have proposed a new non-separable operator and embedded it in the main framework of the cooperation mechanism of bee foraging (see [45] for details).

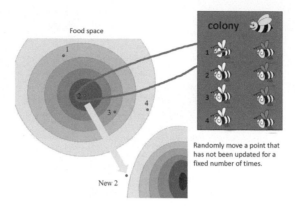

Figure 3.37: Scout bees' search.

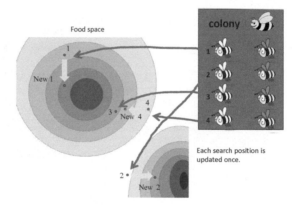

Figure 3.38: Employed bees' new search.

3.4 Firefly Algorithms

Fireflies glow due to a luminous organ and fly around. This glow is meant to attract females. The light generated by each firefly differs depending on the individual insect, and it is considered that they attract others following the rules described below:

- The extent of attractiveness is in proportion to the luminosity.

- Female fireflies are more strongly attracted to males that produce a strong glow.

- Luminosity decreases as a function of distance.

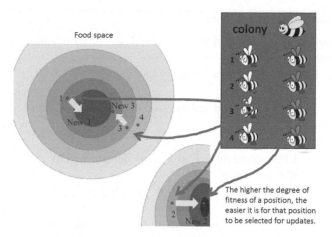

Figure 3.39: Onlooker bees' new search.

The firefly algorithm (FA) is a search method based on blinking fireflies [129]. This algorithm does not discriminate gender. That is, all fireflies are attracted to each other. In this case, the luminosity is determined by an objective function. To solve the minimization problem, fireflies at a lower functional value (with a better adaptability) glow much more strongly. The most glowing firefly moves around at random.

Algorithm 2 describes the outline of the FA. The moving formula for a firefly i attracted by firefly j is as follows:

$$\vec{x}_i^{new} = \vec{x}_i^{old} + \beta_{i,j}(\vec{x}_j - \vec{x}_i^{old}) + \alpha(rand(0,1) - \frac{1}{2}), \qquad (3.35)$$

where $rand(0,1)$ is a uniform random numbers between 0 and 1. α is a parameter to determine the magnitude of the random numbers, and $\beta_{i,j}$ represents how attractive firefly j is to firefly i, i.e.,

$$\beta_{i,j} = \beta_0 e^{-\gamma r_{i,j}^2}. \qquad (3.36)$$

The variable β_0 represents how attractive fireflies are when $r_{i,j} = 0$, which indicates that the two are in the same position. Since $r_{i,j}$ represents the Euclid distance between firefly i and j, their attractiveness varies depending on the distance between them.

The most glowing firefly moves around at random, according to the following formula:

$$\vec{x}_k(t+1) = \vec{x}_k(t) + \alpha(rand(0,1) - \frac{1}{2}) \qquad (3.37)$$

The reason for this is that the entire population converges to the locally best solution in an initial allocation.

Algorithm 2 Firefly algorithm

Initialize a population of fireflies $\vec{x}_i$ $(i = 1, 2, \cdots, n)$ ▷ Minimizing objective function $f(\vec{x})$, $\vec{x} = (x_1, \cdots, x_d)^T$.
Define light absorption coefficient γ
$t = 1$ ▷ Generation count.
while $t < MaxGeneration$ and the stop criterion is not satisfied **do**
    for $i = 1$ to n **do** ▷ for all n fireflies
        for $j = 1$ to n **do** ▷ for all n fireflies
            Light intensity I_i, I_j at $\vec{x}_i, \vec{x}_j$ is determined by f
            if $I_i > I_j$ **then**
                Move firefly i towards j in all d dimensions
            end if
            Attractiveness varies with distance r via $e^{-\gamma r}$
            Evaluate new solutions and update light intensity
        end for
    end for
    Rank the fireflies and find the current best
    $t = t + 1$
end while
Postprocess results and visualization

As the distance becomes greater, the attractiveness becomes weaker. Therefore, under the firefly algorithms, fireflies form groups with each other at a distance instead of gathering at one spot.

The firefly algorithms are suitable for optimization problems on multimodality and are considered to yield better results compared to those obtained using PSO. It has another extension that separates fireflies into two groups and limits the effect on those in the same group. This enables global solutions and local solutions to be searched simultaneously.

3.5 Cuckoo Search

The cuckoo search (CS) [128] is meta–heuristics based on brood parasitic behavior. Brood parasitism is an animal behavior in which an animal depends on a member of another species (or induces this behavior) to sit on its eggs. Some species of cuckoos (Fig. 3.40) are generally known to exhibit this behavior. They leave their eggs in the nests of other species of birds such as the great reed warblers, Siberian meadow buntings, bullheaded shrikes, azure-winged magpies, etc.[1] Before leaving, they demonstrate an interesting behavior, referred to as egg

---

[1]Parasitized species are almost always fixed for each female cuckoo.

Figure 3.40: Cuculus poliocephalus. The lesser cuckoo image from a Japanese stamp.

mimicry: They take out one egg of a host bird (foster parent) already in the nest and lay an egg that mimics the other eggs in the nest, thus, keeping the numbers balanced.[2] This is because a host bird discards an egg when it determines that the laid egg is not its own.

A cuckoo chick has a remarkably large and bright bill; therefore, it is excessively fed by its foster parent. This is referred to as "supernormal stimulus." Furthermore, there is an exposed skin region at the back of the wings with the same color as its bill. When the foster parent carries foods, the chick spreads its wings to make the parent aware of the region. The foster parent mistakes it for its own chicks. Thus, the parent believes that it has more chicks to feed than it actually has and carries more food to the nest. It is considered to be an evolutional strategy for cuckoos to be fed corresponding to their size because a grown cuckoo is many times larger than the host.

The CS models the cuckoos' brood parasitic behavior based on three rules, as described below:

- A cuckoo lays one egg at a time and leaves it in a randomly selected nest.

- The highest quality egg (difficult to be noticed by the host bird) is carried over to the next generation.

- The number of nests is fixed, and a parasitized egg is noticed by a host bird with a certain probability. In this case, the host bird either discards the egg or rebuilds the nest.

---

[2]Furthermore, a cuckoo chick having just been hatched, expels all the eggs of its host. For this reason, a cuckoo chick has a pit in its back to place its host's egg, clamber up inside the nest and throw the egg out of the nest. This behavior was discovered by Edward Jenner, famous for smallpox vaccination.

Algorithm 3 Cuckoo search

Initialize a population of n host nests ▷ Minimizing objective function $f(\vec{x})$, $\vec{x} = (x_1, \cdots, x_d)^T$.

Produce one egg x_i^0 in each host $i = 1, \cdots, n$

$t = 1$ ▷ Generation count.

while $t < MaxGeneration$ and the stop criterion is not satisfied **do**

    Choose a nest i randomly

    Produce a new egg $\vec{x}_i^t$ by performing Lèvy flights ▷ Brood parasite of the cuckoo.

    Choose a nest j randomly and let its egg be $\vec{x}_j^{t-1}$

    if $f(\vec{x}_j^{t-1}) > f(\vec{x}_i^t)$ **then** ▷ The new egg is better.

        Replace j's egg by the new egg, i.e., $\vec{x}_i^t$

    end if

    Sort the nests according to their eggs' performance

    A fraction (p_a) of the worse nests are abandoned and new ones are built by performing Lèvy flights

    $t = t + 1$

end while

Postprocess results and visualization

Algorithm 3 shows the CS algorithm. Based on this algorithm, a cuckoo lays a new egg in a randomly selected nest, according to Lévy flight. This flight presents mostly a short distance random walk with no regularity. However, it sometimes exhibits a long-distance movement. This movement has been identified in several animals and insects. It is considered to be able to represent stochastic fluctuations observed in various natural and physical phenomena, such as flight patterns, feeding behaviors, etc.

Specifically, Lévy distribution is represented by the following probability density function, referred to in Fig. 3.41:

$$f(x; \mu, \sigma) = \begin{cases} \sqrt{\frac{\sigma}{2\pi}} \exp\left[-\frac{\sigma}{2(x-\mu)}\right] (x-\mu)^{-3/2} & (\mu < x), \\ 0 & (\text{otherwise}), \end{cases} \tag{3.38}$$

where μ represents a positional parameter, and σ represents a scale parameter. Based on this distribution, Lévy flight mostly presents a short distance movement, while it also presents a random walk for a long distance movement with a certain probability. For optimization, it facilitates an effective search compared to using random walk (Gaussian flight) according to a regular distribution [128].

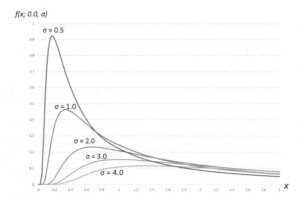

Figure 3.41: Lévy distribution.

Let us consider an objective function represented as $f(\vec{x})$, $\vec{x} = (x_1, \cdots, x_d)^T$. A cuckoo then creates a new solution candidate for the nest i given by the following equation:

$$\vec{x}_i^{t+1} = \vec{x}_i^{t} + \alpha \otimes \text{Lévy}(\lambda), \tag{3.39}$$

where $\alpha(> 0)$ is related to the scale of the problem. In most cases, $\alpha = 1$. The operation $\otimes$ represents multiplication of each element by α. Lévy(λ) represents a random number vector whereby each element follows a Lévy distribution, and this is accomplished as follows:

$$\text{Lévy}(\lambda) \sim rand(0,1) = t^{-\lambda} \quad 1 < \lambda \leq 3, \tag{3.40}$$

where $rand(0,1)$ is a uniform random number between 0 and 1. This formula is essentially a random walk with the distribution achieved by powered steps with a heavy tail. Therefore, it includes infinite averages and infinite standard deviation. An exponential distribution with exponents from -1 to -3 is normally used for a long-distance movement of Lévy flight.

p_a represents a parameter referred to as the switching probability, and its fraction of the worse nests are abandoned from the nest by a host bird and new ones are built by performing Lèvy flights. This probability strikes a balance between exploration and exploitation.

CS is considered to be robust, compared with PSO and ACO [14].

3.6 Harmony Search (HS)

Harmony search (HS) [33] is a meta-heuristic based on jazz session (generation process of human improvisation). Musicians are considered to perform improvisation mainly using any one of the methods as outlined below:

■ Use already-known scales (stored in their memory).

■ Partially change or modify the already known scales. Play scales next to the one stored in their memory.

■ Create new scales. Play random scales within their playable area.

A process whereby musicians combine various scales in their memory for the purpose of composition is regarded as a sort of optimization. While many meta-heuristics are based on swarm intelligence of life, such as fish, insects, etc., HS significantly differs from them in terms of exploiting ideas from musical processes to search for harmony, according to an aesthetic standard.

Harmony search algorithms (referred to as "HS," hereinafter) search for the optimum solution by imitating a musician's processes according to the following three rules:

■ Select an arbitrary value from HS memory.

■ Select a value next to the arbitrary one from HS memory.

■ Select a random value within a selectable range.

With HS, a solution candidate vector is referred to as a harmony, and a set of solution candidates is referred to as a harmony memory (HM). Solution candidates are replaced within HM by a specific order. This process is repeated a certain number of times (or until the conditions for termination are met), and finally, the best harmony is selected among those that survive in HM as a final solution.

Algorithm 4 shows a Harmony Search algorithm. *HMCR* (Harmony Memory Considering Rate) represents the probability of selecting a harmony from HM, while *PAR* (Pitch Adjust Rate) represents the probability of amending a harmony selected from HM. *HMS* is the number of harmonies (sets), which is normally set to between 50 and 100.

A new solution candidate (harmony) is generated from HM based on *HMCR*. *HMCR* is the probability of selecting component elements[3] among the present HM. Thus, new elements are randomly generated by the probability of $1 - HMCR$. Subsequently, mutation occurs according to the probability of *PAR*. The *bw* parameter (Bandwidth) represents the largest size of the mutation. In case a newly generated solution candidate (harmony) is better than the poorest solution of HM, they are replaced.

---

[3]It corresponds to allele in genotype under GA.

Algorithm 4 Harmony search

---

for $i = 1$ to *HMS* **do**   ▷ HM initialization.
    for $j = 1$ to n **do**   ▷ n: harmony length.
        Randomly initialize x_j^i in HM   ▷ x_j^i: j-th position of i-th harmony.
    end for
end for
while the stop criterion is not satisfied **do**▷ Generate a new solution candidate $\vec{x}$.
    for $j = 1$ to n **do**
        if rand(0,1)< *HMCR* **then**
            Let x_j in $\vec{x}$ be the j-th dimension of a randomly selected HM member

            if rand(0,1)< *PAR* **then**
                Apply pitch adjustment distance bw to mutate x_j
                $x_j = x_j \pm \text{rand}(0,1) \times bw$
            end if
        else
            Let x_j in $\vec{x}$ be a random value
        end if
    end for
    Evaluate the fitness of $\vec{x}$ by $f(\vec{x})$
    if $f(\vec{x})$ is better than the fitness of the worst HM member **then**
        Replace the worst HM member with $\vec{x}$ ▷ HM update
    else
        Disregard $\vec{x}$
    end if
end while
Postprocess results and visualization

This method is also similar to a genetic algorithm (GA); however, it differs in that all the members of HM become a parent candidate in HS, while only one or two existing range(s) of chromosomes (parent individual) is/are used to generate a child chromosome in GA.

3.7 Cat Swarm Optimization (CSO)

Cat swarm optimization (CSO) [16] is an optimization method based on a cat's behavior. Cats remain calm most of the time, however, they are interested in their surroundings and are prudent. The length of time spent chasing prey is extremely short, so as to save energy, compared with that of their resting time.

CSO has two main modes:

- **seeking mode**: Represents the resting time (observing the surroundings while taking a rest). Cats make a decision to move when they sense prey and danger.

- **chasing mode**: Represents the time spent chasing prey. After finding a prey, they determine the chasing speed and destination. Thereafter, they start moving.

With CSO, multiple cats are generated in a searching space and each of them becomes a solution candidate. These cats are divided into two groups: Seeking mode and chasing mode. The ratio of these modes is a mixing ratio (*MR*: The number of cats in chasing mode / that in seeking mode). Cats spend most of their time in a calm state in order to observe the surroundings, therefore, *MR* is normally set small.

Seeking mode has four parameters, as listed below:

- *SMP*: Indicates how many cats in seeking mode are considered. It is used to define the memory size of the cats.

- *SRD*: Indicates how much each parameter changes. It is the width of the mutation in a selected dimension. If changed, the difference between the old and new values should be within the range of *SDR*.

- *CDC*: Indicates how many of the parameters are changed. It defines the number of elements to change.

- *SPC*: Indicates whether a cat is a candidate for moving or not. It indicates if the current location of a cat is a candidate for the destination.

Under the condition that *M* represents the number of dimensions in the searching space, the process involved in the seeking mode toward the k–th cat (cat_k) is as follows:

Step1 Set j as follows:

$$j = \begin{cases} SMP - 1 & \text{if } SPC \text{ is true} \\ SMP & \text{otherwise} \end{cases} \qquad (3.41)$$

Step2 Copy the current position of the k–th cat (cat_k) for j times.

Step3 For each copy created, replace the positions based on the following equation:

$$x_{k,d} = (1 \pm SRD \times rand(0,1)) \times x_{k,d}. \tag{3.42}$$

With the condition $d \in \{1,2,\cdots,M\}$, the number of elements to be changed (different numbers of d) is selected within the upper limit of the CDC.

Step4 Calculate the adaptability (fitness) of the entire candidate points (FS_i) under the condition $1 \le i \le j$.

Step5 Calculate the selective probability P_i of each candidate point (cat_i) using the following equation:

$$P_i = \frac{|FS_i - FS_b|}{FS_{max} - FS_{min}} \quad (1 \le i \le j), \tag{3.43}$$

where FS_i represents a functional value (adaptability, i.e., fitness value) of cat_i. FS_{max} and FS_{min} are the maximum and minimum value of the function, respectively. If an objective function is used for searching for the minimum, then $FS_b = FS_{max}$, if it is for the maximum, then $FS_b = FS_{min}$.

Step6 Based on roulette selection with the selection probability P_i, select the destination among the current candidate positions at random, and replace the positions of cat_k.

When in tracing mode, each cat moves at its own speed. This mode is represented by the following three steps:

Step1 Update the velocity ($v_{k,d}$) for each dimension d by following the formula;

$$v_{k,d} = v_{k,d} + rand(0,1) \times c_1 \times (x_{best,d} - x_{k,d}). \tag{3.44}$$

With the condition $d \in \{1,2,\cdots,M\}$, $x_{best,d}$ represents the position of a cat with the best adaptability. c_1 is a user-defined scaling parameter.

Step2 Check whether the velocity exceeds the maximum. If so, update the maximum speed.

Step3 Update the position of each cat, according to the following equation:

$$x_{k,d} = x_{k,d} + v_{k,d}. \tag{3.45}$$

Algorithm 5 Cat swarm optimization

Initialize a population of cats cat_i $(i = 1, 2, \cdots, n)$ with random positions $\vec{x}_i$ in M-dimensional space   ▷ n: the number of cats.
Randomly assign each cat a velocity in range of the maximum velocity value (i.e., $\vec{v}_i; i = 1, 2, \cdots, n$).
while the stop criterion is not satisfied **do**
    According to MR, assign each cat a flag of the seeking or tracing mode
    Calculate the fitness function values for all cats and sort them
    X_g = the best cat   ▷ find the cat with the best solution.
    for $i = 1$ to n **do**
        if cat_i's mode is in seeking mode **then**
            Start seeking mode
        else
            Start tracing mode
        end if
    end for
end while
Postprocess results and visualization

3.8 Meta-heuristics Revisited

Some critical opinions have been expressed against meta-heuristics and further discussions have also been undertaken.

Meta–heuristics frequently uses unusual names, terms associating with nature, and metaphors. For example, in the harmony search, the following terms are used:

■ harmony

■ pitchAnoteA

■ sounds better

However, these are just saying the following words listed below in another way:

■ solution

■ decision variable

■ has a better objective function value

Algorithm 6 Evolution Strategy: $(\mu + 1)ES$

Initialize the population with randomly generated solutions
while the stop criterion is not satisfied **do**
    Create a new solution using recombination and mutation operators
    if the new solution is better than the worst solution in the population **then**
        Replace the worst solution by the new one
    end if
end while
return the best solution in the population

Although critics insist that these replaced words cause confusion [112], it is considered that the use of metaphorical expressions itself does not influence the ease of understanding. For example, David Hilbert[4] is quoted as saying Geometry does work even if a point, line and face are expressed as a table, chair, and beer mug in discussing mathematical forms. That is to say, it does not matter at all in precise discussions when it is axiomatically defined. Nevertheless, we should be careful about insisting on the novelty of meta-heuristics. It is important to recognize the distinct difference with existing methods for further discussions.

Wayland[122] has criticized the Harmony Search as simply a special example of evolution strategy $(\mu + 1)ES$.

Let us recall $(\mu + 1)ES$, one of the algorithms described in section 2.1.2 (see Algorithm 6 in page 95). Comparing this algorithm to that of the harmony search (refer to page 91), it is assumed that the essential method of calculation is the same although they slightly differ in parameter setting, among others [122]. On the contrary, discussion on methods of expanding the harmony search algorithm and the parameter setting methods can be built up in a different way to ES[5].

It does matter that researchers who propose meta-heuristics do not recognize other similar methods [112]. When developing a new method, we never fail to have discussions on the basis of past investigations.

There is a well-known Chinese proverb which states "wēngùzhīxīn" by Confucius[6]. Returning to an old method and expanding it to a produce a new one is not a bad thing and is in fact recommended. Especially, in the pursuit of re-

---

[4]David Hilbert (1862–1943): German mathematician. At the 2nd International Congress of Mathematicians (ICM) in Paris in 1900, he made a speech on "problems in mathematics," where he stressed the importance of 23 unsolved problems and presented a prospect for future creative research through these problems. Some of them continue to be themes for research on mathematics and computer science.

[5]Further discussions have been presented, and there is an insistence on the part of some individuals that it is a different method. Refer to https://en.wikipedia.org/wiki/Harmony_search for details.

[6]Confucius (BC552–BC479): A philosopher in the Spring and Autumn period in China. He is the founder of Confucianism. The "Analects of Confucius"(The Governing Volume) is a collection of sayings and anecdotes attributed to him and his disciples. For over two thousand years, the ideology of Confucius was respected and inherited as China's legitimate ideology and the idea of "learning from the past to develop new ideas" is attributed to the Analects (Wei Chang book). Confucius has supposedly said that a requisite for someone to be his disciple is to study ideas and academic disciplines from his predecessors.

search into rapidly developing AI, long-forgotten technologies are expected to revive with these new technologies (big data, supercomputers, among others), and many examples are well known. Now is the time to recognize the importance of engaging in productive discussions with a particular emphasis on old methods. The expectation in right-minded research communities is to allow various proposed methods to compete against each other, which should result in the continuation of more robust approaches for future generations after a process of successive elimination[7].

---

[7]Those who engage in AI research based on evolution naturally expect evolutionary phenomena (e.g., speciation, diversity, natural selection, preadaptation, punctuated equilibrium and so on) in research results.

Chapter 4

Emergent Properties and Swarm Intelligence

Pristine originality is an illusion; all great ideas were thought and expressed before a conventional founder first proclaimed them. Conventional founders win their just reputations because they prepare for action and grasp the full implication of ideas that predecessors expressed with little appreciation of their revolutionary power.

—Stephen Jay Gould

4.1 Reaction-Diffusion Computing

The Reaction-Diffusion Cellular Automaton is a model of a two-dimensional chemical reaction. In the actual chemical reaction, creation of a certain molecule is accompanied by precipitation of another molecule. When the first molecule is propagated, the reaction stops. A model realizing such phenomenon on a computer is called Reaction-Diffusion Computing.

This computational model is used for image processing. In fact, diffusion can be used for noise filtering and reaction for contrast enhancement. Depending on the parameters, Reaction-Diffusion Computing can realize the following image manipulations:

- restore broken contours

- detect edges

- improve contrast

For example, there is a study [1, 2] to realize the basic operation of image processing using Belousov-Zhabotinsky reaction described in section 5.4.

We will explain the following processes as an application:

■ Generation of the Voronoi diagram

■ Construction of a skeleton from a contour map

4.1.1 *Voronoi Diagram Generation*

Assuming that a set of points (called seeds, sites, or generators) is given. Let us consider constructing the following diagram (called Voronoi diagram; see Fig. 4.2):

■ All the points in the polygon are nearer to the generator within it than to the other generators.

■ For a two-dimensional Euclidean plane, the border of the region becomes part of the bisector of each generator.

A polygon containing a generator is called a Voronoi polygon.

Voronoi polygon is considered as animals' territory in terms of generators. For example, in 2011, a new species of blowfish (Torquigener albomaculosus) was found in Amami isl., Japan (called "Galapagos of the East"). Although the length of a male specimen is just $15cm$, they construct complex radial-shaped nests at the bottom of the sea with $2m$-diameter tunnels (see Fig. 4.1). After observing such nests, divers started to refer to them as "mystery circles." When an egg-laying bed (circle) is completed, a female accepts to lay eggs in the nest for reproduction. A Voronoi polygon is formed when multiple such nests are located close together.

Voronoi diagrams can be seen in nature frequently. Figure 4.3 shows a coral example. These diagrams are used varyingly in engineering as follows:

■ Collision-free path planning

■ Determination of service areas for power substations

■ Nearest-neighbor pattern classification

■ Determination of largest empty figure

The algorithm for creating Voronoi diagrams has been studied in detail in the field of Computational Geometry, but it is considerably complicated and computationally intensive.

When generators $p_1, p_2, p_3, \cdots$, and p_n are given, the simplest method is as follows[1]:

---

[1] In this section, we consider a Voronoi diagram on a two-dimensional Euclidean space.

Figure 4.1: A mystery circle (@Amami isl., Japan in 2019).

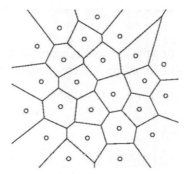

Figure 4.2: A Voronoi diagram.

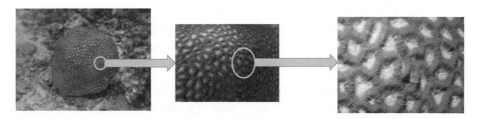

Figure 4.3: A coral example (Favia favus) of Voronoi diagram (@Anilao, Philippines in 2018).

Step1 Draw the perpendicular bisector l_{12} connecting points p_1 and p_2, and find the half plane H_{12} on the p_1 side with the vertical line as the borderline.

Step2 Similarly, obtain half plane H_{13} for p_3 and find the common area of the region $H_{12} \cap H_{13}$.

Step3 Similarly, obtain half plane H_{14} for p_4 and find the common area of the region $H_{12} \cap H_{13} \cap H_{14}$.

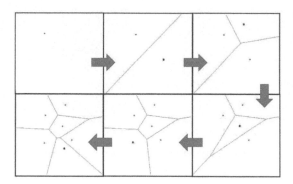

Figure 4.4: Successive Addition Method.

Step4 In the same way, find the area of the common part $H_{12} \cap H_{13} \cap H_{14} \cap \cdots \cap H_{1n}$. This is the Voronoi polygon V_1 for generator p_1.

Step5 The same operation is carried out with respect to other points, yielding the Voronoi polygon V_i $(i = 2, 3, \ldots, n)$ of the generator p_i, and obtain the entire Voronoi diagram.

In the worst case, this algorithm takes the trouble (has a complexity) of $O(n^2)$ to create one Voronoi polygon. Therefore, the computational complexity to create the entire Voronoi diagram, in the worst case, can become $O(n^3)$.

To reduce the computational complexity, the following efficient method has been proposed.

Successive Addition Method First creates a Voronoi diagram with p_1, p_2, p_3 and then adds points $p_4, p_5, \cdots, p_n$ one by one to it. Figure 4.4 shows how a point is added in this process.

Recursive Bisection Method Divide n generators by x coordinates to the left and right with almost $n/2$ each side. This method recursively creates a Voronoi diagram for each side and merges them to create full Voronoi diagram. This method is based on Divide and Conquer paradigm.

Computational time of the successive addition method is $O(n^{3/2})$ when generators are not added in any particular order. However, when the addition order is devised, it becomes $O(n)$ on average. The recursive bisection method on average requires $O(n \log(n))$ computational time.

Now let us create a Voronoi diagram with Reaction-Diffusion Computation. Here, a Voronoi diagram is derived using a Reaction-Diffusion Cell Automaton [1].

In the following, O(n)-Algorithm with $n+3$ states and O(1)-Algorithm with four states are explained.

$O(n)$-Algorithm takes the following $n+3$ states:

Symbols	State
$1,2,\cdots,n$	Excitement (active oxidation)
$\bullet$	Dormant state (neither)
$-$	Refractory (active reduction)
$\#$	Precipitation (bisector creation)

The update rule is as follows:

$$
x^{t+1} = \begin{cases}
\#, & (x^t \in \{1,\cdots,n,\bullet\} \wedge \mid I(x)^t \mid \geq 2) \vee (x^t = \#), \\
+, & (x^t = \bullet) \wedge I(x)^t = \{+\}, \\
-, & (x^t = +) \wedge (I(x)^t = \{+\} \vee I(x)^t = \phi), \\
\bullet, & (x^t = -) \vee (I(x)^t = \phi \wedge (x^t = \bullet)),
\end{cases}
\tag{4.1}
$$

where $1,2,\cdots$, and n are $+$ (i.e., $+ \in \{1,2,\cdots,n\}$). $I(x)^t$ is an excited (+) neigh-boring cell at time t.

As an example, according to this update rule, the following transition occurs:

$$
\begin{aligned}
- &\implies \bullet \\
\# &\implies \# \\
+ &\implies \begin{cases} - & \text{when } \mid I(x)^t \mid = 0 \vee I(x)^t = \{+\} \\ \# & \text{when } \mid I(x)^t \mid \geq 2 \end{cases} \\
\bullet &\implies \begin{cases} \bullet & \text{when } I(x)^t = \phi \\ + & \text{when } I(x)^t = \{+\} \\ \# & \text{when } \mid I(x)^t \mid \geq 2 \end{cases}
\end{aligned}
$$

Figure 4.5 shows a simple evolution example of cellular automaton by $O(n)$-Algorithm in the 4-neighbor case. In this figure, $\bullet$ is a blank cell and $*$ represents a generator.

$O(1)$-Algorithm takes the following 4 states:

Symbols	State
$+$	Excitement (active oxidation)
$\bullet$	Dormant state (neither)
$-$	Refractory (active reduction)
$\#$	Precipitation (bisector creation)

When using $O(1)$-Algorithm, its update rule is as follows.

$$x^{t+1} = \begin{cases} \#, & ((x^t = \bullet) \wedge (S_v(x)^t = 2 \vee S_h(x)^t = 2)) \vee (x^t = \#) \vee ((x^t = +) \wedge (S(x)^t \geq 1)), \\ +, & ((x^t = \bullet) \wedge ((S(x)^t \geq 1) \wedge (S_v(x)^t \neq 2) \wedge (S_h(x)^t \neq 2)), \\ -, & (x^t = +) \wedge (S(x)^t = 0), \\ \bullet, & (x^t = -) \vee ((x^t = \bullet) \wedge (S_v(x)^t < 2) \wedge (S_h(x)^t < 2)). \end{cases}$$

$$(4.2)$$

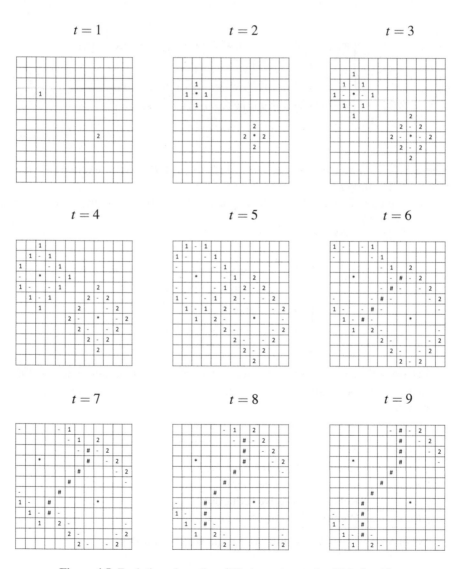

Figure 4.5: Evolution of reaction-diffusion automata by O(n) algorithm.

Where $S(x)^t$ is an excited (+) neighboring cell at time t. $S_h(x)$ and $S_v(x)$ denote the numbers of excited (+) neighbors located horizontally[2] and vertically[3], respectively.

In case of 4 neighbors, transitions are as follows:

$$- \implies \bullet$$

$$\# \implies \#$$

$$+ \implies \begin{cases} - & \text{when } S(x)^t = 0 \\ \# & \text{when } S(x)^t \geq 1 \end{cases}$$

$$\bullet \implies \begin{cases} \bullet & \text{when } S(x)^t = 0 \\ + & \text{when } (S(x)^t \geq 1) \wedge (S_v(x)^t < 2) \wedge (S_h(x)^t < 2) \\ \# & \text{when } (S_v(x)^t = 2) \vee (S_h(x)^t = 2) \end{cases}$$

In case of 8 neighbors, transitions are as follows:

$$- \implies \bullet$$

$$\# \implies \#$$

$$+ \implies \begin{cases} - & \text{when } S(x)^t < 4 \\ \# & \text{when } S(x)^t \geq 4 \end{cases}$$

$$\bullet \implies \begin{cases} \bullet & \text{when } S(x)^t = 0 \\ + & \text{when } (4 > S(x)^t \geq 1) \wedge (\neg P(x)^t) \\ \# & \text{when } (S(x)^t \geq 4) \wedge P(x)^t \end{cases}$$

The predicate $P(t)^t$ is defined as follows:

$$P(t)^t = ((x_N^t = + \wedge x_S^t = +) \vee (x_W^t = + \wedge x_E^t = +) \vee (x_{NW}^t = + \wedge x_{SE}^t = +) \vee (x_{NE}^t = + \wedge x_{SW}^t = +)),$$
(4.3)

where subindices mean North, South, West, East, North-West, South-East, North-East and South-West neighbors of cell x, respectively.

Figure 4.6 shows a simple evolution example of cellular automaton by O(1)-Algorithm.

Figure 4.7 shows the result of the derivation of the Voronoi diagram for two points under various conditions. Let the distance between two points on x axis be dis_x and the distance on y axis be dis_y.

From this figure, we can see that both definition of the neighborhood and the change in distance between the two points affect how region is divided, and the thickness of the division line. In particular, when using O(1)-Algorithm for $dis_x = dis_y$, the area is not completely bisected. In the 4-neighborhood O(N)-Algorithm, the line that divides the area is a vertical bisector when $dis_x \neq dis_y$ and the border does not disappear when $dis_x = dis_y$.

Figure 4.8 shows the Voronoi diagram when 10 points are randomly distributed. The figure shows from the left: The result of O(N)-Algorithm in case

---

[2] East and west neighbors
[3] North and south neighbors

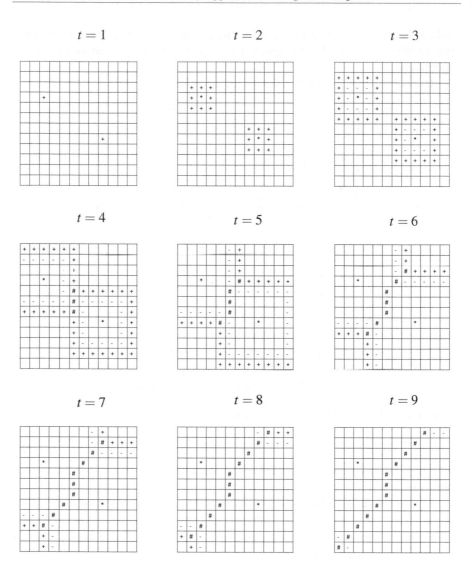

Figure 4.6: Evolution of reaction-diffusion automata by O(1) algorithm.

of 4-neighborhood, the result of O(*N*)-Algorithm in case of 8-neighborhood, the result of O(1)-Algorithm in case of 4-neighborhood, and the result of O(1)-Algorithm in case of 8-neighborhood[4].

Figure 4.9 shows the outlook of Voronoi simulator. This simulator generates Voronoi diagrams when it is provided with a seed point. Each seed is assigned with separate colors, and propagation color is different from each of them. When

---

[4]Moore neighborhood, up, down, left, right and diagonal.

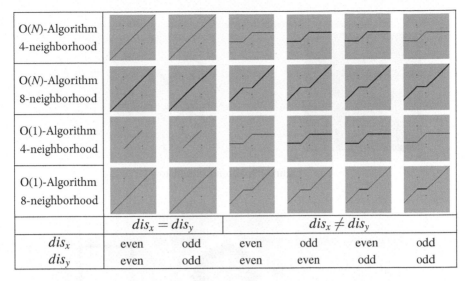

	$dis_x = dis_y$		$dis_x \neq dis_y$			
dis_x	even	odd	even	odd	even	odd
dis_y	even	odd	even	even	odd	odd

Figure 4.7: Voronoi diagram for two points.

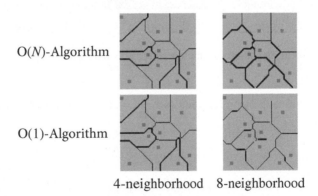

O(N)-Algorithm

O(1)-Algorithm

4-neighborhood   8-neighborhood

Figure 4.8: Voronoi diagram for ten points.

different propagations come into contact, they are designed to change to the color of the boundary. The following shows the experimental results of this simulator.

In the algorithm described so far, a chemical substance propagates to the surrounding area at the next time step. In the previous study [1], L_1 norm and L_∞ norm were used in the experiments as the definition of "surroundings." In addition to that, this simulator is capable of handling the case of the L_2 norm.

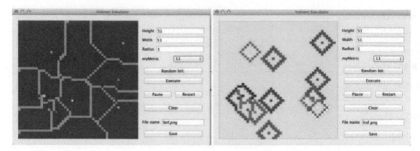

Figure 4.9: Voronoi Simulator.

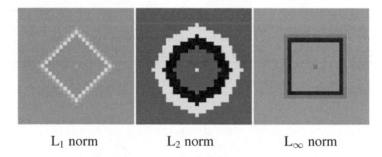

| L_1 norm | L_2 norm | L_∞ norm |

Figure 4.10: The difference in propagation due to the norm (one-point case).

The definition of each norm for vectors $\vec{x} = (x_1, x_2, \cdots, x_n)$ is as follows:

$$
\begin{aligned}
L_1 \text{norm} \quad &: \quad |x_1| + |x_2| + \cdots + |x_n| \\
L_2 \text{norm} \quad &: \quad \sqrt{x_1^2 + x_2^2 + \cdots + x_n^2} \\
L_\infty \text{norm} \quad &: \quad \max\{|x_1|, |x_2|, \cdots, |x_n|\}
\end{aligned}
$$

Here, the radii of the L_1 norm and the L_∞ norm are 1 and the radius of the L_2 norm is 3. Since the field is discrete, the L_1 norm and L_2 norm are the same for radius of 1 and 2. The difference in propagation due to the norm is shown in Fig. 4.10.

Let us see the difference for each norm of the boundary for two seed points. As can be seen in Fig. 4.11, if two seed points are parallel to the axes, the shape of the boundary does not depend on the seeds. That is, the line connecting two points becomes a perpendicular bisector. This is because the L_1, L_2 and L_∞ norms match on the axis. Figure 4.12 shows the propagation for two seeds in a more general positional relationship. In this case, the difference of the boundary due to the differences in the norms was observed. It can be seen that the boundary due to the L_1 norm and boundary due to L_∞ norm are mirror images (in the mirror-image relationship).

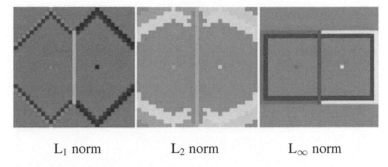

L_1 norm $\qquad$ L_2 norm $\qquad$ L_∞ norm

Figure 4.11: The difference in propagation due to the norm (two-point case).

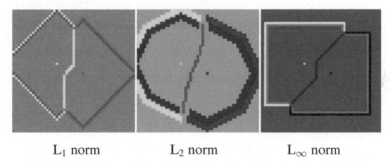

L_1 norm $\qquad$ L_2 norm $\qquad$ L_∞ norm

Figure 4.12: The difference in propagation due to the norm (general two-point case).

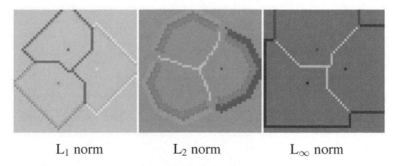

L_1 norm $\qquad$ L_2 norm $\qquad$ L_∞ norm

Figure 4.13: The difference in propagation due to the norm (three-point case).

Figure 4.13 shows the difference between the boundaries of each norm due to three seed points. A trend similar to the two-point case is seen, that is, L_1 and L_∞ are in a mirror-image relationship. To be more precise, it is rotated 90 degrees around the center of mass of the three seeds and is in reversed axis relationship. In other words, the boundary of L_1 and the boundary of L_∞ are in the folding

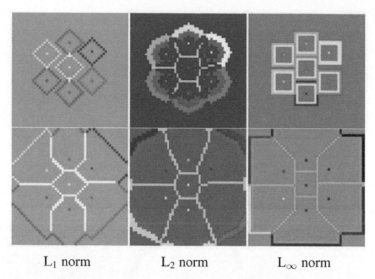

L₁ norm   L₂ norm   L∞ norm

Figure 4.14: The difference in propagation due to the norm (hexagon case).

Color version at the end of the book

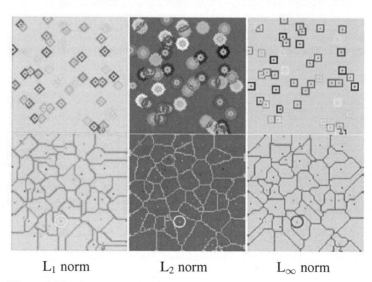

L₁ norm   L₂ norm   L∞ norm

Figure 4.15: The difference in propagation due to the norm (random case).

relationship. The axis of the folding is the axis for the L_2. However, it should be noted that it is a discrete field, and there is a slight deviation due to the thinning.

Figure 4.14 shows the result when seed points are arranged in a hexagon shape. The upper row shows the development process and the lower row shows the case just before the end. For L_1 the central part is hexagonal, for L_2 it is a circle and for L_∞ it is a square.

Figure 4.15 shows the state of the simulation when many random seed points were scattered. It can be seen that the results confirmed so far appeared in various places.

As is evident from the above results, the generated Voronoi diagram is quite different depending on the norm. For example, paying attention to the three Voronoi diagrams in the lower row in Fig. 4.15, the differences become apparent. For each norm in the region of the seed, there is a clear difference in shape and area. When considering the Voronoi diagram as a simulation of a real physical phenomenon, it is necessary to pay close attention to the selection of the norm.

4.1.2 Thinning and Skeletonization for Image Understanding

Thinning is a technique that is often used as a pre-processing operation in the fields of character recognition and pattern recognition. The process of converting a binary image into a 1-pixel wide image is called thinning (see Fig. 4.16).

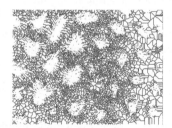

Figure 4.16: Thinning a coral picture.

Various thinning algorithms are known. For example, Hilditch's thinning [49] is executed as follows:

Step1 Let the background pixel be 0 and the figure pixel be 1.

Step2 Search for pixels that satisfy the following condition.

- ■ It is a figure pixel (1).
- ■ It is a boundary point.
- ■ It does not delete endpoints.
- ■ It preserves isolated points.
- ■ It preserves connectivity.

Step3 Erase pixels satisfying the above conditions.

Step4 Repeat steps from **Step**1 to **Step**3 until there are no more pixels to remove.

For another example, Tamura's thinning algorithm [115] uses the pattern shown in Table 4.1 and is executed as follows:

Step1 In cases of pattern no.1, remove the pixel (i.e., change the color of the center from black to white). If it corresponds to an exceptional pattern, do not remove.

Step2 If no pixel is removed, end the process. If some pixel is removed, proceed to the next step.

Step3 In caes of pattern no.2, remove the pixel. If it corresponds to an exceptional pattern, do not remove.

Step4 If no pixel isremoved, end the process. If some pixel is removed, go back to **Step**1.

Note that a cell of 1 stands for a black pixel, 0 for a white pixel, and $*$ for either black or white in the table.

Although the conventional algorithms are complicated, it is possible to easily realize thinning operation using Reaction-Diffusion CA. This method is basically the same as O(n)-Algorithm and uses the following four states.

Symbol	State
EXC	Excitement (active oxidation)
REF	Refractory (active reduction)
RES	Dormant state (neither)
PRE	Precipitation (bisector creation)

Assuming that the number of excited (EXC) neighboring cells of the cell x at time t is $s^t(x)$, the update rule of the thinning operation is as follows (in case of 8 neighbors):

$$
x^{t+1} = \begin{cases}
\text{EXC}, & (x^t = \text{RES}) \wedge (1 \leq s^t(x) \leq 4), \\
\text{REF}, & (x^t = \text{EXC}) \wedge (s^t(x) \leq 4) \vee (x^t = \text{REF}), \\
\text{RES}, & (x^t = \text{RES}) \wedge (s^t(x) = 0), \\
\text{PRE}, & \{(x^t = \text{RES}) \vee (x^t = \text{EXC})\} \wedge (s^t(x) \geq 5) \vee (x^t = \text{PRE}),
\end{cases}
$$

$$(4.4)$$

Figure 4.17 shows the result of the application of the thinning. Results were achieved for different thicknesses of the bar T. The black lines in the green area are the result of the thinning operation. A successful case of the applied thinning operations can be seen in the left-most figure. Note that changing the thickness of the bar can make blank spaces between the lines or sometimes even unintended points. A more complicated example is given in Fig. 4.18.

Table 4.1: Patterns used for Tamura's algorithm.

Patterns no. 1

Patterns to be removed	$\begin{array}{ccc} & 1 & \\ & 0 & \\ & & \end{array}$	$\begin{array}{ccc} & & \\ & 0 & 1 \\ & & \end{array}$

| Exceptional Patterns (Part I) | $\begin{array}{ccc} & 1 & \\ 0 & 0 & \\ 0 & 1 & \end{array}$ | $\begin{array}{ccc} 1 & 0 & \\ 0 & 0 & 1 \\ & & \end{array}$ |

Exceptional Patterns (Part II) Same as patterns no. 2	$\begin{array}{ccc} 1 & 0 & 1 \\ & & \\ & 0 & \end{array}$	$\begin{array}{ccc} & 1 & \\ 0 & & 0 \\ & 1 & \end{array}$	$\begin{array}{ccc} & 0 & \\ 1 & 0 & 1 \\ & & \end{array}$	$\begin{array}{ccc} & 1 & \\ & 0 & 0 \\ & 1 & \end{array}$	$\begin{array}{ccc} 1 & 0 & \\ & 0 & 1 \\ & & \end{array}$	$\begin{array}{ccc} 0 & 1 & \\ 1 & 0 & \\ & & \end{array}$
	$\begin{array}{ccc} 1 & 0 & \\ & 0 & 1 \\ & & \end{array}$	$\begin{array}{ccc} & & \\ & 0 & 1 \\ & 1 & 0 \end{array}$	$\begin{array}{ccc} 1 & 0 & 1 \\ 0 & 0 & 0 \\ 1 & & 1 \end{array}$	$\begin{array}{ccc} 1 & 0 & 1 \\ & 0 & 0 \\ 1 & 0 & 1 \end{array}$	$\begin{array}{ccc} 1 & & 1 \\ 0 & 0 & 0 \\ 1 & 0 & 1 \end{array}$	$\begin{array}{ccc} 1 & 0 & 1 \\ & 0 & 0 \\ 1 & 0 & 1 \end{array}$

Patterns no. 2

Patterns to be removed	$\begin{array}{ccc} & 0 & \\ & 1 & \\ & & \end{array}$	$\begin{array}{ccc} 1 & 0 & \\ & & \\ & & \end{array}$

| Exceptional Patterns (Part I) | $\begin{array}{ccc} 1 & 0 & \\ 0 & 0 & \\ & 1 & \end{array}$ | $\begin{array}{ccc} & & \\ 1 & 0 & 0 \\ & 0 & 1 \end{array}$ |

Exceptional Patterns (Part II) Same as patterns no. 1	$\begin{array}{ccc} 1 & 0 & 1 \\ & & \\ & 0 & \end{array}$	$\begin{array}{ccc} & 1 & \\ 0 & & 0 \\ & 1 & \end{array}$	$\begin{array}{ccc} & 0 & \\ 1 & 0 & 1 \\ & & \end{array}$	$\begin{array}{ccc} & 1 & \\ & 0 & 0 \\ & 1 & \end{array}$	$\begin{array}{ccc} 1 & 0 & \\ & 0 & 1 \\ & & \end{array}$	$\begin{array}{ccc} 0 & 1 & \\ 1 & 0 & \\ & & \end{array}$
	$\begin{array}{ccc} 1 & 0 & \\ & 0 & 1 \\ & & \end{array}$	$\begin{array}{ccc} & & \\ & 0 & 1 \\ & 1 & 0 \end{array}$	$\begin{array}{ccc} 1 & 0 & 1 \\ 0 & 0 & 0 \\ 1 & & 1 \end{array}$	$\begin{array}{ccc} 1 & 0 & 1 \\ & 0 & 0 \\ 1 & 0 & 1 \end{array}$	$\begin{array}{ccc} 1 & & 1 \\ 0 & 0 & 0 \\ 1 & 0 & 1 \end{array}$	$\begin{array}{ccc} 1 & 0 & 1 \\ & 0 & 0 \\ 1 & 0 & 1 \end{array}$

Figure 4.17: Thinning process by reaction-diffusion computing.

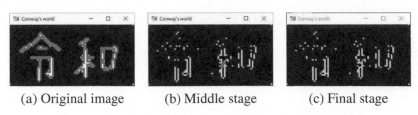

(a) Original image   (b) Middle stage   (c) Final stage

Figure 4.18: Thinning process by reaction-diffusion computing.

4.2 Queuing Theory and Traffic Jams

4.2.1 *Most Random Customers and Poisson Arrival*

When does the wait occur? Let us model a customer arriving at random (most irresponsible customers). Dice and roulette can be named as the representatives of the randomness.

The randomness of the dice includes three elements:

Independence The next face that comes out does not depend on the faces that have come out so far. It does not depend on the past.

Stationarity The probability of the face being shown does not depend on time. Dice does not wear down.

Binarity There are six possible outcomes, from 1 to 6, with no intermediate values.

Independence is also said to be memorylessness.

Considering the most random customer from analogy to the dice, it becomes as follows:

Independence The probability that the customer will come in unit time is the same on average.

Stationarity The probability of the arrival within a certain time does not depend on when that time begins.

Scarcity Two people will not come in a minute time (its probability is zero).

Binarity was replaced by scarcity.

As is well known, the probability that after rolling the dice n times, the face "1" comes out k times is:

$$_nC_k \left(\frac{1}{6}\right)^k \left(1 - \frac{1}{6}\right)^{n-k} \tag{4.5}$$

In this way, an experiment in which trials are repeated independently, an event occurs with probability p and does not occur with $1 - p$, is called a Bernoulli trial. The probability of an event occurring k times after repeating the trial n times is:

$$_nC_k p^k (1 - p)^{n-k} \tag{4.6}$$

If the arrival of the random customer is thought of as an independent trial (judging whether or not customers arrive within Δt time), the probability that k people will come within the time t is the probability of a Bernoulli trial of an event occurring k times within n trials. Note that the probability p that customers will arrive at each interval is:

$$p = \lambda \cdot \Delta t = \lambda \cdot \frac{t}{n}, \tag{4.7}$$

and λ is the average arrival rate.

Definition 4.1   The number of people arriving within the unit time is called the average arrival rate λ, and its unit is the number of people/hour.

Here, if the value n is made sufficiently large, the probability that the number of people arriving at time t is k becomes the Poisson distribution of λt.

Theorem 4.1
The discrete probability distribution of the arrival number (per unit time) of the average arrival rate of λ [people/hour] is described by Poisson distribution:

$$P(N = k) = \frac{\lambda^k}{k!} e^{-\lambda}. \tag{4.8}$$

Therefore, the arrival number of the most random customers becomes a Poisson distribution and is called Poisson arrival.

Figure 4.19 shows an example of Poisson distribution. Poisson distribution has the following properties:

■ The average value and the variance are equal (λ).

■ As the average value increases,

    ■ Distribution becomes symmetric.

    ■ It approaches normal distribution.

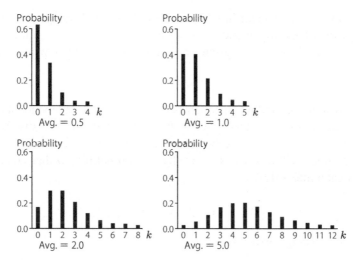

Figure 4.19: Poisson distribution.

The stochastic process based on this distribution is modeled for a series of random events and is called a Poisson process.

The density distribution of the time (arrival time interval) from the arrival of one customer until the arrival of another customer is the exponential distribution of the average $1/\lambda$.

Theorem 4.2
The arrival time distribution of the average arrival time of $1/\lambda$ (time/person) is an exponential distribution expressed as:

$$f(t) = \lambda e^{-\lambda t}. \tag{4.9}$$

Figure 4.20 shows an example of exponential distribution. The exponential distribution has the feature that both average and standard deviation are equal to $1/\lambda$. Also, it is important to show the memorylessness[5] explained in page 112.

---

[5]Stochastic variable X conforming to the exponential distribution satisfies the following equation:

$$P(X > s+t \mid X > s) = P(X > t), \quad where \; s,t \geq 0. \tag{4.10}$$

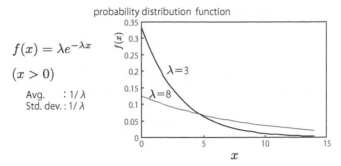

Figure 4.20: Exponential distribution.

The fact that the arrival time interval is an exponential distribution, is well suited to the situation that "when it comes it keeps coming, and when it does not come then it does not come easily." In other words, the less time it takes for the next customer to come, the more exponentially it grows.

4.2.2 Poisson Distribution and Cognitive Errors

Consider the following as a familiar problem of probability:

Assume that in the town where you live, there is a possibility of a lightning strike at any time of the year. The frequency is about once a month. In other words, the probability of lightning striking per day is about 0.03. Today on Monday, lightning fell in your town. Then, which one of the following has the highest probability of *the next* lightning strike? Answer with a reason.

(a) Tuesday tomorrow

(b) A month later

(c) Probability on any day is the same

(d) None of the above (write answers specifically)

This is a typical example of the Poisson process.

A summary of answers to this problem is shown in Table 4.2[6]. Surprisingly, most of the students chose the answer (c) that the probability is the same for

---

[6]Questionnaire survey from the lecture on "artificial intelligence" at the author's university. Lecture for 3rd-year undergraduate students. Each year about 100 people attend it. Participants are mainly from engineering departments, with students from literature, law, and medical departments.

Table 4.2: Comparison of the ratio of answers.

	Probability of lightning strike	
	Frequency	(Answer)
Correct	16.5%	(Tuesday tomorrow)
Incorrect	67.4%	(Probability on any day is the same)
	11.6%	(A month later)
	2.0%	(Otherwise)

every day, and only about 17% of the students chose the correct answer (a), even though students majoring in science were the main subjects of the questionnaire.

The reason as to why the correct answer is (a) "Tomorrow" is as follows. The probability that the next lightning strike will be tomorrow (Tuesday) is 0.03. On the other hand, the probability that the next lightning strike will be the day after tomorrow (Wednesday) is $0.03 \times 0.97 = 0.0291$, because it corresponds to a situation that the lightning strike happens on Wednesday and it does not fall tomorrow on Tuesday. Likewise, the probability that the next lightning strike will be on Thursday is $0.03 \times 0.97 \times 0.97 = 0.0282$. In this manner, the probability decreases (exponentially) with every passing day. However, there are not many people who understand it correctly. Despite changing the font and underlining the word "next" to avoid overlooking, the answer "the probability on any day is the same" is most common. Some answers were correct (a), but some students answered with the wrong reason that "bad weather is likely to appear in series." There is a similar example with a hurricane. The frequency for a hurricane happening once in 100 years, is not important. Hurricane of this class hit Florida two consecutive years from 2004, causing serious damage. As a result, the insurance company (Poe Financial Group) lost 10 years worth of the surplus money and bankrupted [31].

Steven Pinker,[7] an evolutionary psychologist who conducted an internet experiment, reported that 5 out of 100 solved the problem of lightning strikes correctly [93].

As mentioned in the previous section, the interval between events in the Poisson process is exponentially distributed. There are many short intervals, but the longer intervals are less common. In other words, events that occur randomly appear to be clustered.

However, people cannot easily understand this probability law. It is said that mathematician William Feller was the first to point out this cognitive illusion. Feller gave an example of London bombing during World War II as an example of the cognitive error [93]. London citizens felt that German V2 rockets re-

---

[7]Steven Arthur Pinker (1954–): American cognitive psychologist. He is a Harvard University Professor of psychology and the author of a large number of scientific publications including "How the Mind Works," "The Blank State: The Modern Denial of Human Nature," and "The Language Instinct: How the Mind Creates Language."

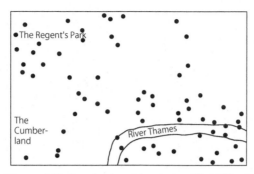

Figure 4.21: London bombing during World War II.

Table 4.3: Conformity to the Poisson law (Flying-bomb attack on London).

#. of flying bombs per square	Expected #. of squares (Poisson)	Actual #. of squares
0	226.74	229
1	211.39	211
2	98.54	93
3	30.62	35
4	7.14	7
5 and over	1.57	1
Total	576.00	576

peatedly hit a district while other districts were not damaged at all. From that, they believed that the German army was capable of targeting particular areas. Figure 4.21 shows the impact areas of 67 V2 rockets that fell in the center of London. Looking at that, it is true that the damage on the upper right and lower left is small compared to the upper left and lower right, where the damage is bigger. This makes it questionable whether a region in the city is safe, or whether the bomb's landing points form a cluster. In other words, is it not randomly distributed throughout London?

To solve this problem, it is best to do statistical tests, but it can also be compared with Poisson distribution. Here, 144 square kilometers of southern London will be divided into 576 parcels ($= 24 \times 24$) each being a square of 1/4 square kilometers. Then, classifying the parcels according to the number of bombs that fell within the period is as shown in Table 4.3 (a total of 537 shots). This table shows expected values of impacts in the parcels (based on the Poisson distribution of the average $\lambda = \frac{537}{576}$) and measured values. From the comparison of the columns of expected and measured values, it can be seen that they are quite consistent. In other words, the distribution of this data is consistent with the Poisson distribution, and the bomb wants to be dropped at random [17].

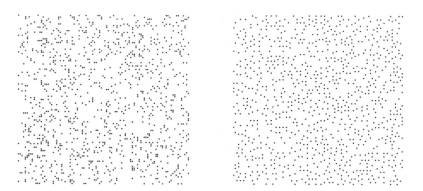

Figure 4.22: Simulation of point generation in two-dimensions: Which is random?

Let us introduce an interesting example pointed out by Stephen Jay Gould [8]. Since ancient times, the human race has been interested in the alignment of stars, gave the names to the constellations, and sometimes felt the mysterious power in conglomerations of stars [42]. Is the sequence of stars random, or is there any intention behind it?

Let us see Fig. 4.22. One of them is a diagram generated by random simulation. Which one looks like a sequence of stars? Most people feel that the left is a sequence of stars and that the right is a random sequence. Actually the truth is opposite. The figure randomly generated by Poisson distribution is the left figure of Fig. 4.22 (1,100 points were generated on 100×100 grid). On the other hand, The right figure is a model of carefully placed lightning bugs (or fireflies, see also sec. 3.4), so that each point does not come close in consideration of each territory.

In this way, humans see some rules in random models, at the same time mistakenly recognizing randomness in artificial models.

In other words, humans cannot easily understand the randomness of Poisson distribution, and they misunderstand that random events are made of clusters and regularity. To create equal intervals, a non-random process is required. As a result, mistakes, such as seen in Table 4.2, are made. This also leads to criticism that true human intelligence (strong artificial intelligence) cannot be realized. This is because the main technologies used in current artificial intelligence and machine learning are strongly dependent on stochastic reasoning and statistics.

---

[8] Stephen Jay Gould (1941-2002): American paleontologist and evolutionary biologist. He advocated the punctuated equilibrium theory of evolutionary process. He wrote an essay every month for American magazine "Natural History," and many books that summarized it became best sellers. As a researcher of the same evolutionary theory, Richard Dawkins (see the footnote in page 185) was his opponent in debate.

Daniel Kahneman[9] and Amos Tversky[10] proposed the following two ideas for such human cognitive errors.

■ Representative heuristics: It is easy to overestimate the probability of the items similar to typical examples.

■ Law of small numbers: In statistics, the great law of approaching the theoretical value by increasing the number of trials. It is easy to think that humans can get close to the theoretical values even with a small number of trials, and a psychological bias is applied.

Gambler's fallacy will be considered as a representative heuristic. Which is easier to get after rolling a fair dice 7 times, 1111111 or 5462135? Stochastically thinking, both are the same (see memorylessness on page 112). However, when numbers are arranged like 5462135, it is regarded as a representative example of randomness[11]. On the other hand, Bayes' Law can justify this error. It depends on strengthening the prior probability that in 1111111 it is easier to roll 1 (it is a cheating dice). 5462135 has no such surprises.

In this way, it seems that reports such as winning lottery tickets being continually sold at the same counter, and random playback of music players playing the same song[12], become more convincing.

4.2.3   Queue Management and Scheduling

Depending on the customer's flow, the queue is formed as follows (see Fig. 4.23).

1. When customers arrive according to arrival distribution, they enter the vacant counter to receive service. The service is provided in a first-in, first-out (FIFO) order.

---

[9] Daniel Kahneman (1934–): Israel-born American psychologist. He proposed the behavioral economics and the prospect theory, integrating psychology and cognitive science, and received the 2002 Nobel Prize in Economics.

[10] Amos Tversky (1937–1996): Israel-born psychologist. He created the prospect theory in a joint research with Kahneman.

[11] However, if you ask this question as follows, the answer will change. "Yesterday I rolled the dice 7 times. Those numbers are written on this paper. Which one is it?" Also, the expected number of times to roll the dice to get 1111111 is $6^7 + 6^6 + \cdots + 6$, whereas the expected value of 5462135 is $6^7 + 6$. This difference in the number of times may be the cause of the error.

[12] In February 2015, Spotify, internet based music distribution service, was accused of the suspicion that song selection of the shuffle play was biased. See also http://www.dailymail.co.uk/sciencetech/article-2960934/Is-music-player-really-random-Spotify-says-users-convinced-strange-patterns-\ \shuffle-playlists.html

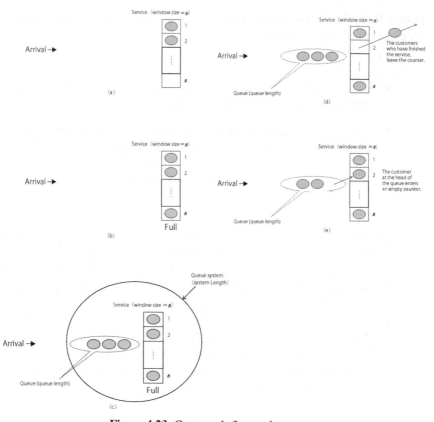

Figure 4.23: Customer's flow and queue.

2. When customers arrive one after another, the service counters get full.

3. Next customers cannot enter the counter, so they make a queue and wait.

4. When customers come one after another, the queue is extended.

5. The customers who have received the service, leave the counter.

6. The customer at the head of the FIFO queue, enters an empty counter and receives service.

7. Repeat the above process.

The basic elements that make up the queue in this way are as follows:

■ Number of counters (Window size)

■ Customer arrival distribution

■ Service time distribution

What is often used is, arrival of the customers according to the most random arrival (Poisson arrival) and the distribution of service time defined as an exponential distribution (that is, the number of people leaving is Poisson distribution).

By modeling like this, it is possible to solve the phenomenon of the queue analytically under some conditions. This is based on the derivation of the equilibrium state (steady distribution), which looks at the stable conditions of the system. Using it allows the system designer to control the queue. Since many queuing models cannot be solved analytically, queue simulators[13] are used. For example, the management strategy of a food shop, where queues are created, can be designed based on labor costs and customer trends.

In addition, it is possible to handle queues more efficiently using scheduling theory. Let us consider doing jobs accumulated in the queue one at a time. In other words, when multiple jobs (tasks) have to be executed, it is a matter of determining the order in which one person (or one CPU) will work

In this case, the priority rule (dispatching rule) of job assignment becomes important. This rule specifies how to resolve conflicts when they occur (i.e., conflict resolution). A conflict means that multiple executable jobs simultaneously request the job to start (competing). Typical dispatching rules include the following:

- ■ SPT (shortest processing time)
  Choose the job with the shortest processing time. This will shorten the average delivery delay, but will also generate jobs with large delivery delay.

- ■ EDD (earliest due date)
  Choose the job closest to the due date. It is valid for the evaluation with respect to the delivery delay.

- ■ SLACK (SLACK time)
  Choose the job with the smallest SLACK time (= delivery date - current time - remaining processing time). Valid for evaluation with respect to the deadlines.

- ■ RANDOM
  Execute the task at random. Although it seems inefficient at first sight, some degree of performance is guaranteed.

- ■ FIFO (first-in, first-out)
  A neutral rule. It has the same characteristics as RANDOM, but there is less variation in the result than in RANDOM.

---

[13]The queue simulator developed by the author can be obtained from the homepage http://www.iba.t.u-tokyo.ac.jp/iba/SE/mimura/howto/HowTo.html

Performance of the scheduling depends on what kind of priority rule is adopted. The difficulty of scheduling lies in:

■ Randomness of the arrival of each task

■ Randomness of processing time of each task

The author has taught at university and instructed many students, and they always have multiple reports with varying degrees of difficulty and deadlines. In some cases, it may not be known beforehand when an assignment will be issued (Poisson arrivals). If the report assignment is not processed appropriately, the queue of reports to be submitted will become longer, and one day it will fail, and reports will not be submitted. As expected, many students take an EDD strategy. However, this strategy is not necessarily good. Although the SLACK strategy seems to be relatively efficient, it is necessary to estimate the "estimated processing time" correctly. There is a heuristics of scheduling, "Do it immediately" (It does not have to be 100%, but do it ASAP). This means the importance of at least trying a little bit, regardless of the deadline, when given the tasks. Then it becomes possible to estimate the processing time beforehand, which makes it possible to take a SLACK strategy or use more general scheduling method.

In order to verify this, a 10-semester simulation was conducted and the seriousness of the students was compared (the time spent on reports per day) with the results (the number of credits acquired) in each strategy. Looking at the results, it is difficult for negligent students[14] to use the most effective scheduling, and it is best to digest the problem with the SPT (Fig. 4.24(a)). It is more efficient to prioritize some of the subjects by giving up on others from the beginning, than trying a little bit in all of them and failing. Also, even for excellent students, the difference due to dispatching rules will not increase (Fig. 4.24(c)). If a student works diligently every day in his own way, he will be able to score well. This may be natural in some ways.

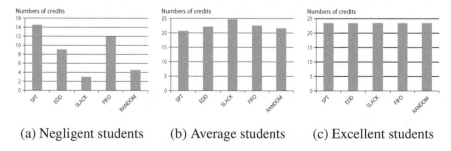

(a) Negligent students   (b) Average students   (c) Excellent students

Figure 4.24: Numbers of credits acquired in each strategy.

---

[14]Diligence is defined as hard work at home.

On the other hand, efficient scheduling is meaningful for the majority of average students (Fig. 4.24(b)). It can be seen that the SLACK strategy is relatively good. SLACK will not be an optimal strategy if one of the tasks is extremely difficult. In heavy tasks, the term "remaining processing time" in SLACK is increased, and other tasks are neglected. As a result of this, the overall grades may become worse. However, in SLACK, the rate at which tasks can be submitted before the deadline is increased. Therefore, it is a relatively load resistant strategy even compared to other strategies.

It was found out that the strategy that does not take anything into consideration, such as RANDOM, works effectively. Interestingly, sometimes it is better than the SPT or EDD. In the fields of artificial intelligence and optimization, random operations may be successful. For example, UCT search for Monte Carlo trees, which recently has drawn attention as a search strategy in games (see page 30) and a random selection algorithm are typical examples.

As for the scheduling methods described above, it has been shown that the emergent approach, such as GA, GP, PSO and ACO, etc., is very effective. In fact, an efficient method based on evolutionary computation has been proposed to approximately solve NP-complete problem, called the JSSP (job scheduling problem,indexproblem!job scheduling problem see [53, ch.2.8.2] for details), in a reasonable time, and was applied to a variety of practical problems.

There is a method called ramp metering, which is used to counter traffic congestion and is often seen in the United States these days. In this method, a traffic signal is placed in front of the accelerating lane entering the expressway to control the inflowing cars (Fig. 4.25). The traffic light alternates between red and blue. For the most part, it is red, and once every few seconds it becomes blue for a moment and only one car is allowed to enter at that moment. It is known that introducing a ramp meter improves traffic congestion. In fact, stopping the ramp meter in Minnesota caused it to look like the following [31].

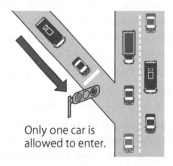

Only one car is allowed to enter.

Figure 4.25: Ramp metering.

- Traffic at peak time was decreased by 9%.

- The required time was increased by 22% and the reliability declined.

- The speed of the car dropped by 7%.

- Collision accidents during confluence sharply increased by 26%.

However, according to a subjective survey of the drivers, it was found out that they were not necessarily satisfied. The reason is that waiting for the ramp meter, even though the actual total time is shortened, makes drivers feel as if they did not get to the destination quickly. It would be better to run, even slowly, than to stop at a ramp meter. In other words, drivers do not like to be kept waiting, and they feel frustrated because they are not in control.

In Disneyland, the following two methods were introduced to efficiently control the queue:

- Fast Pass: Go to the venue near the attraction beforehand and obtain the fast pass. You can enter the attraction without waiting, during the time zone specified on the fast pass. Merit for the customers using fast pass is the reduction of time spent in queues for popular attractions, while Disneyland reduces the variance of the arrival time. The number of fast passes issued in a certain time period is fixed.

- The indication of the long expected wait time: The customer is able to ride more attractions than was originally planned by making a schedule with a long-expected waiting time, thus, the satisfaction level is increased[15].

According to the real experience, since admissions using fast pass are overwhelmingly less in numbers than the general admissions, thus, there seems to be no dissatisfaction that ordinary customers were overtaken at the time of the meeting. In addition, since the number of times fast pass can be used is limited, and while waiting for the arranged time zone, customers wait in queues for other attractions and in total the number of visited attractions does not differ so much between the customers. Therefore, customers are less dissatisfied, and the variation of the arrival rate is relaxed, which is considered a great advantage.

Even if the "long wait time" is shorter than the actual time, the customer feels more satisfied. On the other hand, it seems that there are many customers who give up the attraction they wanted to ride and go home because they were afraid of waiting too long for their turn. In this case, unnecessary complaints

---

[15] Another theory is that the longer display is unintentional. It is due to measurement error by the Flick Card.

are generated from customers. It is also said to reduce the motivation of the customers to line up for thrilling vehicles[16].

There is a popular website "touringplans.com" to create an itinerary to go around more attractions, with as few wait times for Disneyland and Disney World[17]. There are many enthusiastic fans of this page. For example, in 2003, Edward Waller and Yvette Bendeck have achieved the feat of conquering all the attractions of the Magic Kingdom in less than 11 hours[18].

On this homepage, the itinerary was created manually by combining queuing theory and the method of assignment problems by linear programming. This is based on the queuing theory, as described in section 4.2.3. It is said that around 37% more of the attractions were visited by using the proposed tour plan. Because the limitations of this approach are seen, the original method has been extended in order to estimate the waiting time from the past data and recent trends so as to solve the traveling salesman problem[19] with evolutionary computation. This successfully reduced the wait time by about 90 minutes compared to the past. Of course, it is also essential to collect good data to create a tour plan. In fact, for more than several years, the investigators have been placed in a theme park to collect data, such as expected wait time. In addition, many members of touringplans have participated in this data collection.

The example described in this section merely suggests the limitation of controlling the queue with optimal criterion optimization. A more flexible model incorporating human sensibility will be necessary.

4.3 Silicon Traffic and Rule 184

Rule 184[20] is known as the Burgers cellular automaton (BCA), and has the following important characteristics (Table 4.4):

1. The number of 1s is conserved.

2. Complex changes in the 0-1 patterns are initially observed, however, a pattern always converges to a right-shifting or left-shifting pattern.

---

[16]https://www.mouseplanet.com/11037/How_Posted_Wait_Times_Compare_To_Your_Actual_Wait_In_Line

[17]https://touringplans.com/,https://www.wired.com/2012/11/len-testa-math-vacation/

[18]https://touringplans.com/walt-disney-world/touring-plans/ultimate/hall-of-fame

[19]Traveling Salesman Problem (TSP):When there are several cities located on the map, the problem of asking for the minimum length of the closed path (called Hamiltonian closed path) that returns to the original by passing through a city once. It is known as a typical NP-complete problem, and there are many applications to circuit wiring and delivery. See [53, ch.2.8.1] for details.

[20]Refer to page 14 for this decimal representation of one-dimensional cellular automata.

Table 4.4: Rule 184.

$a_t^{i-1}, a_t^i, a_t^{i+1}$	111	110	101	100	011	010	001	000
a_{t+1}^i	1	0	1	1	1	0	0	0

(a) $\alpha = 1.0$,no signal　　(b) $\alpha = 0.8$,no signal　　(c) $\alpha = 0.8$,signal type 110

Figure 4.26: Congestion simulation using BCA.

Take a_t^i as the number of cars on lattice point i at time t. The number is 1 (car exists) or 0 (no car exists). The car on lattice point i moves to the right when the lattice point is unoccupied, and stays on the lattice point when occupied at the next time step. This interpretation allows a simple description of traffic congestion using rule 184.

BCA can be expanded as follows:

■ BCA with signals: Restrict traffic to the right at each lattice point

■ Probabilistic BCA: Move with a probability (α)

Figure 4.26 shows a result of BCA simulation. The top row shows the current road conditions, and the time passed as we move down. Cars are shown in red in this simulator. Continuous red areas represent congestion because cars cannot move forward. Congestion is more likely to form when the probability α that a car moves forward is small. (a) shows how congestion forms (red bands wider than two lattice points form where a car leaves the front of the congestion and another car joins at the rear). A signal is added at the center in (c). The signal pattern is shown in blue when 1 and red when 0, and the pattern in this simulation was 110 (i.e., blue, blue, red, blue, blue, red,···).

A simulation of silicon traffic based on these models is shown in Fig. 4.26. A two-dimensional map is randomly generated, and cars are also placed at random. Two points are randomly chosen from the nodes on the map, and are designated as the origin and the destination. The path from the origin to the destination is determined by Dijkstra's algorithm. Here, a cost calculation, where the distances between nodes are weighted with the density of cars in each edge, is performed in order to determine a path that avoids congestion. The signal patterns are designed to change the passable edge in a given order at a constant time interval.

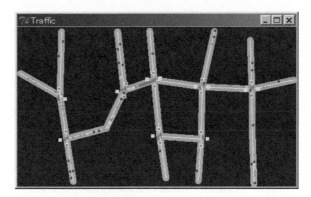

Figure 4.27: Simulation of silicon traffic.

The following is another explanation of traffic modeling. Nagel et al. modeled traffic congestion using a one-dimensional CA [80]. In their model, each cell corresponds to a position where cars can pass, and each cell can take one of two states (one or no car is in the cell) in every time step. Every car moves at a characteristic speed (integer value from 0 to v_{max}). The speed of the cars and state of the cells are updated based on the following rules:

Acceleration Increase the speed by one unit ($v := v + 1$) if the speed of a car v is less than v_{max} and the distance from the first car ahead is greater than $v + 1$.

Deceleration Decrease the speed of a car at i to $j - 1$ ($v = j - 1$) if the first car ahead is located at $i + j$ and $j \leq v$.

Random number Decrease the speed of all cars by 1 (if larger than 0) at possibility p ($v := v - 1$).

Move Move all cars at their respective speed v.

The most important parameter in this simulation is the density of cars ρ defined as

$$\rho = \frac{\text{total number of cars}}{\text{total number of cells}} \tag{4.11}$$

Figure 4.28 is an example of the flow of cars with $v_{max} = 5$ and $\rho = 0.1$. Cars move from left to right, and the right end is connected to the left end. The numbers indicate the speed of the cars, and the dots show cells with no cars. Consecutive cars at speed 0 at the center indicate traffic congestion.

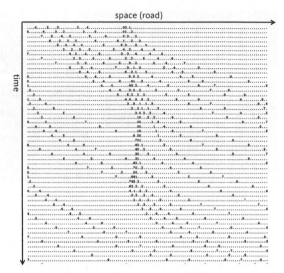

Figure 4.28: An example of the flow of cars.

Nagel et al. conducted various experiments and found that the nature of congestion changes at $\rho = 0.08$. Note that congestion is measured by the average speed of all cars. Thresholds such as $\rho = 0.08$ are called critical values and are important parameters in the simulation of complex systems.

Nagels et al. performed the modeling of traffic jams using a one-dimensional lattice. In their model, each cell corresponds to the vehicles' passage, at each time-step, the cells take one of the two states (vacant or occupied by one vehicle). All the vehicles move with a specific speed (integer number, 0 to v_{max}). For example, the movement of a car with $v_{max} = 10, \rho = 0.1$ is shown in Fig. 4.29. The vehicle moves in the right-hand direction, and the right end and the left end are connected. The top row shows the current road conditions, and the time passed as we move down (displayed till 200 steps). The vehicles' speed is represented by shades of colors. In fact, green becomes darker with the increase in speed, and the red becomes darker as the speed decreases. A chunk of red color is the point of occurrence of a traffic jam, and the black part shows a cell without a vehicle. In Fig. 4.29, the occurrence of a series of vehicles with speed 0 in the middle is a traffic jam. We can see that the traffic jam moves forward with time.

In this model, SlowtoStart has been introduced to realize the effect of inertia. This is a rule that says, once the vehicle has stopped, then it starts moving after 1 time step even if the front is open to move. This is considered important to bring the model closer to the actual stability. If SIS is used, it takes time to accelerate, which leads to more stationary vehicles and worsens the traffic jam (Fig. 4.29). On the other hand, without SIS, the line of the stationary vehicles will not elongate unless there is a slowdown due to a random number, and, as a result, the traffic jam is eased (Fig. 4.30).

Figure 4.29: Simulating traffic jam (with SIS).

Color version at the end of the book

Figure 4.30: Simulating traffic jam (without SIS).

Color version at the end of the book

Recently, ASEP (Asymmetric Simple Exclusion Process) is being studied extensively as a model of traffic flow [96]. In this model, the maximum speed of each vehicle existing in each cell is taken as 1, if the cell in front is vacant, the vehicle moves to it with a probability p (stops with a probability $1 - p$). Let us take the inflow and outflow probabilities of new cars as α and β, respectively (in other words, right and left ends are not connected, and the number of vehicles is not fixed). Figure 4.31 shows the ASEP model simulation. Parameters are $\alpha = 0.3, \beta = 0.9, p = 0.5$ for (a) free phase, $\alpha = 0.3, \beta = 0.3, p = 0.5$ for (b) shockwave phase, and $\alpha = 0.6, \beta = 0.6, p = 0.5$ for (c) maximum flow phase. As explained earlier, the topmost row is the current road situation. As we go down, it shows the time already passed. In the ASEP model, various mathematical analyses are done for the equilibrium state [96].

Southeast Asia has many lanes with a dead end. They come together to form a main line for traveling back or passing one another. Main lines usually have some U-turn points for traveling to the opposite side. However, making a U-turn is not immediately possible because the main line is large. Furthermore, traffic is usually backed up for about an hour due to the heavy traffic volume of the line.

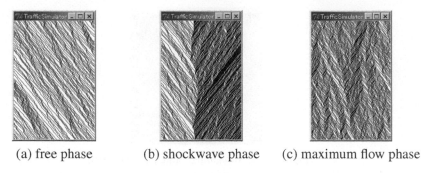

(a) free phase   (b) shockwave phase   (c) maximum flow phase

Figure 4.31: ASEP model simulation.

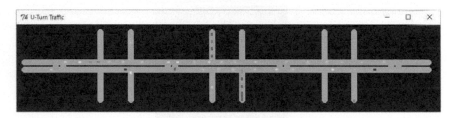

Figure 4.32: Main line with U-turn places.

Thus, main lines must have U-turn points, however, these installations have some trade-offs.

■ With fewer U-turn points, the cost of cars running in the wrong direction would increase.

■ Too many U-turn points would generate a traffic jam because cars making U-turns would hinder the traffic.

Considering this optimization, we shall perform a simulation experiment. Figure 4.32 shows a model of a road in Southeast Asia. It comprises one-way lanes and a long main line. The two-way main line has U-turn points so that vehicles traveling along the one-way lane can move to another place. One-way lanes have either exit only (out edge) or enter only (in edge).

The experiment described here was performed with several conditions. In the simulation described above, it was considered that a vehicle's movement was only the distance between the vehicle in front and the intersection of a red signal, and an angle at the signal was excluded. To make a natural U-turn, the vehicle slows down at the U-turn point, turns to the vertical direction and waits until sufficient space is gained in the opposite lane. Accordingly, we performed simulations with the conditions set as follows:

■ The following one-dimensional equation of motion is used to estimate the necessary distance (S_1) for the car to slow down:

$$S_1 = \frac{v_{now}^2 - v_{turn}^2}{2a} \times \frac{2\theta}{\pi}. \qquad (4.12)$$

In this formula, v_{now} and a represents the current velocity and acceleration of the vehicle, respectively. θ represents an angle toward the destination[21]. v_{turn} represents an appropriate velocity for making a U-turn and should be 20% of the maximum velocity. The distance S_1 decreases according to θ, i.e., the direction toward the destination. If S_1 is smaller than the distance to the U-turn point, the vehicle should make the U-turn.

■ The following formula determines the distance for a vehicle for the minimum distance to the same U-turn point among those heading for it in the opposite lane to stop there:

$$S_2 = \frac{v_{now}^2}{2a} \qquad (4.13)$$

If S_2 is shorter than the actual distance to the U-turn point, the vehicle avoids making the U-turn because it cannot safely stop.

■ Any vehicle must clear the way for an ambulance to pass by. Therefore, the acceleration should be decreased to half.

As previously stated, since one-way lanes are either exit only (out edge) or enter only (in edge), selecting any one from each of the six out edges and six in edges generates 924 combinations. Moreover, the main lines can be selected in two ways, which generates 1,848 combinations for selection. For these combinations, we performed the simulation with 100 running vehicles for a traffic rate of 0.02. Figure 4.33 shows the plot between the number of U-turns made and the actual time taken to make the U-turn. It indicates that the more clever the decision making process for the U-turn with repetitions, the shorter the time to reach the destination (the averaged delay time). Nevertheless, this result was made under limited conditions, therefore, further detailed data analysis is required.

The same traffic simulation experiments have been conducted in a two-dimensional map (see Fig. 4.34 for example).

4.4 Segregation and Immigration: What is Right?

The two different colored pieces (referred to as agents below) are arranged in two-dimensional space. In that space, each agent wants its neighborhood (Moore

---

[21] $0 \le \theta \le \frac{\pi}{2}$ and θ becomes closer to 0 as the destination is behind. In the case that the destination is somewhere after making a turn and moving straight ahead, $\theta = \frac{\pi}{2}$.

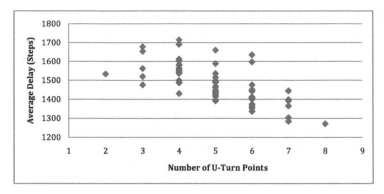

Figure 4.33: The result of U-Turn road experiment.

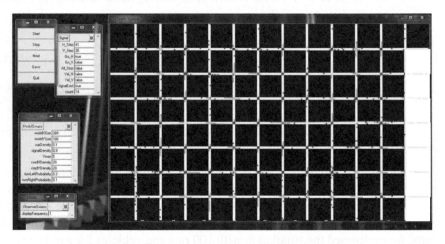

Figure 4.34: Traffic simulation with a two-dimensional map.

neighborhood, 8 neighbors up, down, left, right and diagonal) to have the same color as itself. In other words, if the colors of the surrounding pieces are different from itself, it moves from there.

This is Thomas Schelling's segregation model [107]. Schelling[22] has studied how macro-level phenomena emerge from micro-level preferences. Here, preference is the threshold percentage of how many neighbors should be of the same color. The following are the basic rules of Schelling's model.

---

[22]Thomas Schelling (1921–2016): American social scientist and economist. Awarded Nobel Prize in Economics in 2015.

Figure 4.35: Thomas Schelling's segregation model. The original experiments were conducted with 23 pennies and 22 dimes.

- All agents belong to one of two groups, and have a preference value on neighbors of the same color.

- Agents calculate the ratio of neighbors of the same color.

- The agent stays if the ratio is above the preference value. Otherwise, the agent randomly moves to an empty space where the preference criterion is satisfied.

In Schelling's era, computers were not familiar, so he placed coins (penny and dime currency) on the chess board, rolled a die and manually experimented (see Fig. 4.35).

Figures 4.36–4.38 show actual experiments following the above rules. In the initial state in which the two groups (i.e., tribes of different colors) were alternately intertwined, when the preference was set to 50%, the group became clearly divided and stabilized (Fig. 4.36). On the other hand, if the preference was set to be smaller than 33%, it stabilized in the initial state. The white piece in the figure indicates vacant land. If the preference was set to be 50% for three group case, then the population became completely isolated (Fig. 4.37). On the other hand, if the preference was 33%, a part of the group was stabilized while being surrounded by other groups (Fig. 4.38). Also, the number of iterations until reaching stability is considerably smaller when the preference was 33%.

Schelling's results can be summarized as below:

1. Two colors segregated when the preference value exceeded 0.33. Thus, the critical value for this simulation is 0.33.

2. When the preference of one color was set higher than that of the other color, the color with the low preference spread out while the color with the high preference came together.

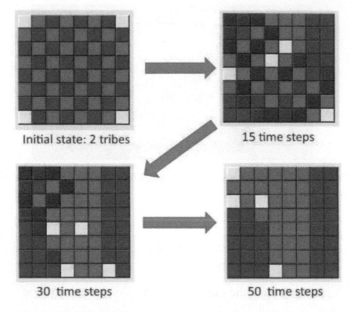

Figure 4.36: Simulation of segregation (2 tribes, 50% preference).

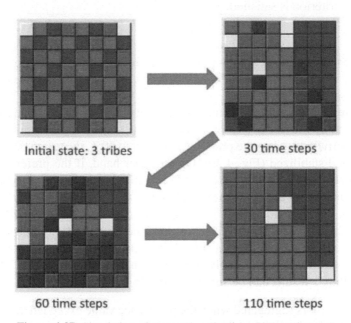

Figure 4.37: Simulation of segregation (3 tribes, 50% preference).

Color version at the end of the book

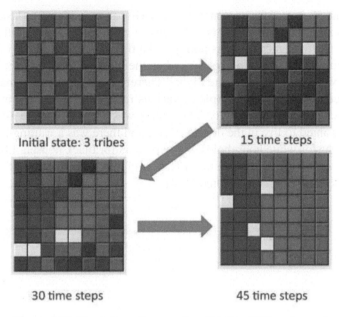

Initial state: 3 tribes   15 time steps

30 time steps   45 time steps

Figure 4.38: Simulation of segregation (3 tribes, 33% preference).

3. The two colors also segregated when the preference criteria was changed to "the agent stays when there are three or more neighbors of the same color."

In summary, two colors ultimately segregated, and all agents wanted all neighbors to be of the same color. Schelling also found that the addition of relatively small changes to the preferences resulted in drastic changes to the macroscopic segregation patterns. In particular, the "prejudice", or bias in preference, of each agent and the segregation pattern is correlated. "Color blind" (pacifists with preference criteria of zero) agents act to mix the colors. However, many critical values exist for these effects. The most significant finding is that society as a whole moves toward segregation even when agents only slightly dislike the other color (low preference threshold). Schelling's results highlighted many problems regarding racial discrimination and prejudice, and extensive research using improved versions of this model is still being conducted.

Schelling's results have raised many problems on "segregation" and "prejudice." A related issue is affirmative action (policies that actively counter discrimination to benefit under-represented groups based on historic and social environmental reasons).

An example is Starrett City in Brooklyn, NY, the largest federally assisted public rental housing scheme in the USA for middle-income residents that started construction in the mid-1970s. There was a requirement for residents that "limited the number of African-American and Hispanic residents to less than 40% of the total number of residents" to create a community where different races co-exist. The reason behind this policy is the theory of the "critical value" in the segregation model. Races do not mix when the ratio of whites becomes less than a certain value because of "white flight." Therefore, an attempt was made to build a stable community where people of various races live by keeping an appropriate balance of race and ethnicity. This policy was a success in one aspect: Many families wanted to live in the development, and there was a waiting list (three to four months for white households and two years for black households).

However, whether these policies (e.g. preferential admission of minorities) are "righteous" is fiercely debated [104].

Chapter 5

Complex Adaptive Systems

All life is problem solving. I have often said that from the amoeba to Einstein there is only one step.

—Karl Popper

5.1 Diffusion-Limited Aggregation (DLA)

Diffusion-Limited Aggregation (DLA) is a growth model of crystals. Growth processes of the following examples can be handled:

- Confetti candy,
- Metal Tree (tree-like metal crystal),
- Bacterial Colony (see section 6.3),
- Snowflake.

These crystals form a fractal, in which its part and the whole structure are similar. This structure can be simulated statistically in DLA.

DLA treats the floating and diffusion of particles in a solvent as a random walk. Those particles are adsorbed on the crystallized part. In addition, melting of the crystal can happen in real phenomena, which is not considered here. In more detail, the DLA behaves as follows:

■ There is a seed crystal in the center of the system.

■ Crystal particles from afar approach with a random walk while also being a subject to Brownian motion.

■ When particles come into contact with a part already in crystal form, it is adsorbed in that place with a certain probability.

■ One particle at a time is a subject to Brownian motion.

In DLA, the growth of the crystal is characterized by varying the adsorption rate when particles are adjacent to the crystal. In other words, when the adsorption rate is high, the particles are more likely to be adsorbed to the outer crystals, thus, crystals will grow thin and long. On the other hand, when the adsorption rate is low, the probability of the particles entering the interior of the crystal becomes high, thus, crystal grows thick.

5.2   How do Snowflakes Form?

Snowflakes have fascinated many people. Japanese scientist Nakaya Ukichiro (1900–1962) conducted research on snow at Hokkaido University. He took microscope photographs of not only beautiful crystals but also variously shaped snow and classified the crystals. He also studied the effect of the weather conditions on the type of falling crystals. Then he created a low-temperature chamber, in which he conducted his research and in 1936, for the first time in the world, he succeeded at creating artificial snowflakes. As a result of this experiment, it was found that changing the temperature and amount of water vapor produces different shaped crystals. Based on that, a famous relation chart called "Nakaya diagram" was created. He mentions the famous words "snow is a letter sent from heaven."

Crystal, depending on its structures, has directions in which it grows more easily, and other directions where the growth is scarce. Crystals are made of regularly arranged atoms and molecules. On top of that, the surface where atoms are densely arranged, becomes a surface that is easier to grow. The water molecules constitute snow crystals. Since water molecules have the property of polarity, as is well known, oxygen molecules are not only covalently bonded to two hydrogen atoms but also loosely bonded to hydrogen atoms of other water molecules (hydrogen bonds). As a result, one oxygen atom is bonded to four hydrogen atoms around it so as to form a tetrahedron. Therefore, the bond angles between all oxygens and hydrogens become 120°. As a result, the oxygen molecules in the solid water crystals are arranged like regular hexagons in a row. Thus, the snow crystals tend to grow in hexagonal symmetry direction.

Adsorption rate= 0.05   Adsorption rate= 0.5   Adsorption rate= 0.99

Figure 5.1: Snowflake growth (1).

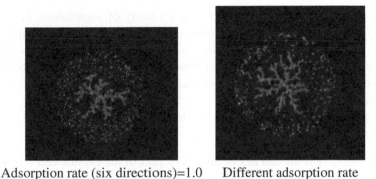

Adsorption rate (six directions)=1.0   Different adsorption rate

Figure 5.2: Snowflake growth (2).

First, the snow crystal was reproduced based on a simple DLA. Anisotropy, as described below, was not considered in this case. Figure 5.1 shows the crystal growth when the adsorption rate was set to be 0.05, 0.9 and 0.99. As can be seen from the figure, when the adsorption rate is low, the crystal grows thick and grows thin when the rate is high. The self-similarity was acquired, however, it did not resemble the snowflake. As it can be seen, the hexagonal symmetry as described above was not acquired.

Next, in order to consider the anisotropy, the adsorption rate of 1 was set when extending to one of the six directions (up right, down right, up left, down left, up, down) and for other directions, it was set to 0.5. The result is presented in Fig. 5.2(a). It certainly extends in those six directions, but it is far from an actual snowflake. In addition, Fig. 5.2(b) shows the case when the adsorption rate was changed according to the number of branches. Although the self-similarity was obtained by this, it did not become like snow crystals. One of the problems is that the simulation was realized in a square lattice shape. For more accurate simulation, it may be necessary to use a hexagonal lattice shape.

Let us simulate a snowflake that is actually closer to the real one. Here its process will be explained based on the Mesoscopic lattice map. Gravner

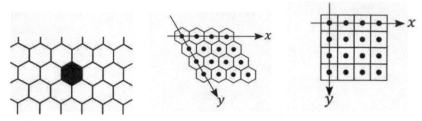

(a) Honeycomb structure   (b) Oblique coordinate   (c) Cartesian coordinate

Figure 5.3: Snowflake coordinate.

et al. [40, 41] succeeded in generating a variety of snowflakes. The space for snowflake formation is a honeycomb structure in which regular hexagons are spread as shown in Fig. 5.3(a). The black hexagon is the constituent unit of the structure (called a cell). Originally it should be displayed in oblique coordinate as shown in the figure (b), but it is re-displayed correctly in the Cartesian coordinate system (c).

The state at the time t of the cell x in the Mesoscopic lattice map is described by the following four values:

■ $a_t(x)$: 1 if the cell at x at time t is a part of the crystal, and 0 otherwise.

■ $b_t(x)$: boundary mass (parameter of a liquid), i.e., the liquified amount in the cell at x at time t

■ $c_t(x)$: crystal mass (parameter of a solid), i.e., the solidified amount in the cell at x at time t

■ $d_t(x)$: diffusive mass (parameter of a gas), i.e., the amount of water vapor in the cell at x at time t

In a state with a seed of the crystal in the center, the initial values are as follows:

Only one cell in the center	Other points
$a_0(x) = 1$	$a_0(x) = 0$
$b_0(x) = 0$	$b_0(x) = 0$
$c_0(x) = 1$	$c_0(x) = 0$
$d_0(x) = 0$	$d_0(x) = \rho$

Where ρ is the amount of water vapor. These four values are updated according to the following four processes per unit time:

- Diffusion: Update $d_t(x)$ of the cell that has not yet crystallized using the neighboring cells' diffusive masses.

- Freezing: Update $b_t(x), c_t(x)$ and $d_t(x)$ in the cell.

- Attachment: The cell is crystallized according to $b_t(x), d_t(x)$ and the number of crystallized neighboring cells.

- Melting: Update $b_t(x), c_t(x)$ and $d_t(x)$ in the cell.

The details of each process are described below. For that purpose the following symbols are defined:

$$
\begin{aligned}
N_x &= \{x\} \cup \{y \mid y \text{ are } x\text{'s neighboring cells}\} \\
A_t &= \{x \mid a_t(x) = 1\} : \text{A set of cells that are crystals at time } t \\
\partial A_t &= \{x \notin A_t \mid a_t(y) = 1, \exists y \in N_x\} : \\
& \quad \text{A set of non-crystal cells in contact with some crystal} \\
A_t^c &= \{x \mid a_t(x) = 0\} : A_t\text{'s complementary set} \\
n_t(x) &= \#\{y \in N_x \mid a_t(y) = 1\} : \text{The number of crystals in cells adjacent to cell } x
\end{aligned}
$$

Note that $|N_x| = 7$ in a honeycomb structure (see Fig. 5.3).

- Diffusion

  For $x \in A_t^c$, update $d_t(x)$ as follows:

$$
d_t(x) \Longleftarrow \frac{1}{7} \sum_{y \in N_x} d_t(y). \tag{5.1}
$$

  For $y \in A_t$, use $d_t(x)$ instead of $d_t(y)$.

- Freezing

  For $x \in \partial A_t$, update $b_t(x), c_t(x)$ and $d_t(x)$ as follows:

$$
\begin{aligned}
b_t(x) &\Longleftarrow b_t(x) + (1 - \kappa) \cdot d_t(x), \tag{5.2} \\
c_t(x) &\Longleftarrow c_t(x) + \kappa \cdot d_t(x), \tag{5.3} \\
d_t(x) &\Longleftarrow d_t(x). \tag{5.4}
\end{aligned}
$$

- Attachment

  For $x \in \partial A_t$, update $a_t(x)$ as follows:

  - If $n_t(x) = 1$ or 2 and $b_t(x) \geq \beta$ then $a_t(x) \Longleftarrow 1$
    If $n_t(x) \geq 3$ and ($b_t(x) > 1$ or ($\sum_{y \in N_x} d_t(y) < \theta$ and $b_t(x) \geq \alpha$)), then $a_t(x) \Longleftarrow 1$

■ If $n_t(x) \geq 4$, then $a_t(x) \Longleftarrow 1$

If $a_t(x)$ is assigned to be 1, update $b_t(x), c_t(x)$ with the following equations:

$$b_t(x) \quad \Longleftarrow \quad b_t(x), \tag{5.5}$$
$$c_t(x) \quad \Longleftarrow \quad b_t(x) + c_t(x). \tag{5.6}$$

■ Melting
  For $x \in \partial A_t$, update $b_t(x), c_t(x)$ and $d_t(x)$ as follows:

$$b_t(x) \quad \Longleftarrow \quad (1 - \mu) \cdot b_t(x), \tag{5.7}$$
$$c_t(x) \quad \Longleftarrow \quad (1 - \gamma) \cdot c_t(x), \tag{5.8}$$
$$d_t(x) \quad \Longleftarrow \quad d_t(x) + \mu \cdot b_t(x) + \gamma \cdot c_t(x). \tag{5.9}$$

Figure 5.4 shows an overview of the snowflake simulator according to the above method. The brief explanation of the parameters used in this simulator is given in Table 5.1.

Figures 5.5–5.7 gives the simulation results repeated up to 10,000 times using the parameters in Table 5.2.

Figure 5.8 shows a simulator of the growth process of the snowflakes. The growth of snowflakes can be observed by setting parameters $\rho, \beta, \alpha, \kappa, \mu, \gamma$ and σ in the bar. Figure 5.9 shows snowflakes generated with various parameters. The simulation parameters in Fig. 5.9(e) are as follows from left to right:

■ $\rho = 0.66, \beta = 1.6, \alpha = 0.4, \theta = 0.025, \kappa = 0.075, \mu = 0.015, \gamma = 0.00005, \sigma = 0.000006$

■ $\rho = 0.65, \beta = 1.6, \alpha = 0.2, \theta = 0.0245, \kappa = 0.1, \mu = 0.015, \gamma = 0.00005, \sigma = 0.00001$

■ $\rho = 0.8, \beta = 2.6, \alpha = 0.006, \theta = 0.005, \kappa = 0.05, \mu = 0.015, \gamma = 0.0001, \sigma = 0.00005$

For more precise simulation, let us visualize how three-dimensional snow crystals are generated. Here, the amount of water vapor is colored and displayed so that the process of snowflake formation can be seen. By doing so, the water vapor is visualized in the areas where the water vapor is scarce.

The three-dimensional visualization of the crystal structure was based on the literature [41]. In this model, calculations are carried out on a lattice with triangular prisms, in which the cell width is 1 in the horizontal direction and $\sqrt{3}$ in the height direction, as shown in Fig. 5.10. In this simulation, the points in the height direction are collectively represented as one cell.

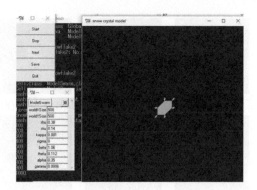

Figure 5.4: Snowflake simulator (1).

Table 5.1: Parameters for snowflake growth.

ρ	vapor density
β	attachment threshold (anisotropy)
α	knife-edge instability (the closer to 0, the more the instability in concavities between ridges)
θ	knife-edge instability (the closer to 0, the more delay in attachment off the ridges)
κ	crystallization (freezing rate$= 1 - \kappa$)
μ	melting rate (boundary mass back to vapor)
γ	melting rate (nonattached ice reverting to vapor)
σ	random fluctuation (noise)

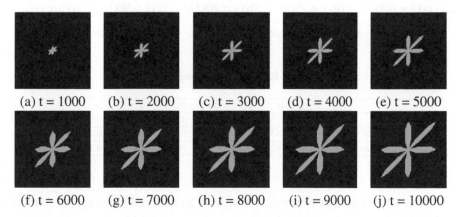

(a) t = 1000 (b) t = 2000 (c) t = 3000 (d) t = 4000 (e) t = 5000

(f) t = 6000 (g) t = 7000 (h) t = 8000 (i) t = 9000 (j) t = 10000

Figure 5.5: Snowflake growth (Simple star).

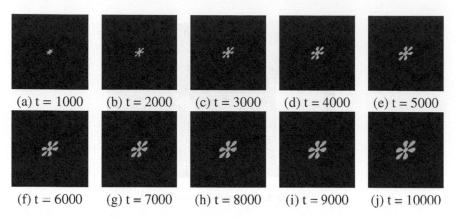

(a) t = 1000　　(b) t = 2000　　(c) t = 3000　　(d) t = 4000　　(e) t = 5000

(f) t = 6000　　(g) t = 7000　　(h) t = 8000　　(i) t = 9000　　(j) t = 10000

Figure 5.6: Snowflake growth (Stellar crystal with plate ends).

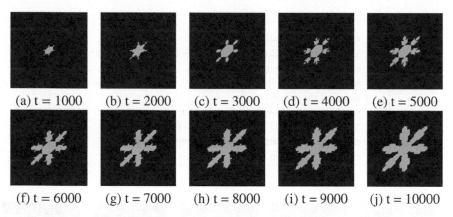

(a) t = 1000　　(b) t = 2000　　(c) t = 3000　　(d) t = 4000　　(e) t = 5000

(f) t = 6000　　(g) t = 7000　　(h) t = 8000　　(i) t = 9000　　(j) t = 10000

Figure 5.7: Snowflake growth (Plate with dendrite ends).

Table 5.2: Parameters for snowflake growth.

	ρ	β	α	θ	κ	μ	γ	σ
(a) Simple star	0.65	1.75	0.2	0.026	0.15	0.015	0.0001	0
(b) Stellar crystal with plate ends	0.36	1.09	0.01	0.0745	0.0001	0.14	0.00001	0
(c) Plate with dendrite ends	0.38	1.06	0.35	0.112	0.001	0.14	0.0006	0

The following three parameters were used in each cell in the crystal formation of the snow:

■　$a(z)$: Indicates whether the z-th layer is frozen.

■　$b(z)$: The amount that is liquified/solidified in the z-th layer.

■　$d(z)$: The amount of water vapor in the z-th layer.

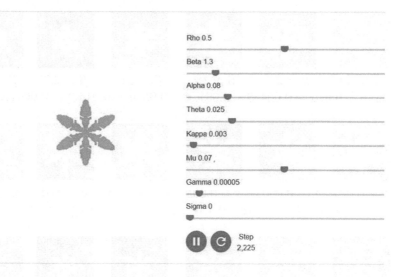

Figure 5.8: Snowflake simulator (2).

In order to reduce the amount of computation, structures with symmetry in the vertical direction and only the single layer were considered, such as star-shaped snow and snow plates. Therefore, the following restrictions were added:

■ If a certain point is frozen, then all the inner layers are frozen.

■ The crystal is formed symmetrically with respect to the central plane.

■ The layer of ice does not form two or more separate layers.

As described above, each point in the vertical direction is calculated as a single cell.

By the first and last assumptions, the points during freezing are limited to the unfrozen bottom of each cell. Thus, the a value (a parameter indicating whether it is frozen) and the b value (a parameter indicating the mass during freezing) can be updated with a single top surface value of the crystal. In addition, it is normally necessary to perform the calculations in both planes with respect to the simulation plane. However, by the second assumption, the amount of calculation can be halved, by obtaining the result for both sides at the same time. Therefore, the parameters of each cell can be reduced as follows:

■ a: Number of consecutive frozen layers from the center plane

■ b: The amount that is liquified/solidified in the $\pm a$-th layer

■ $d(z)$: The amount of water vapor in the $\pm z$-th layer

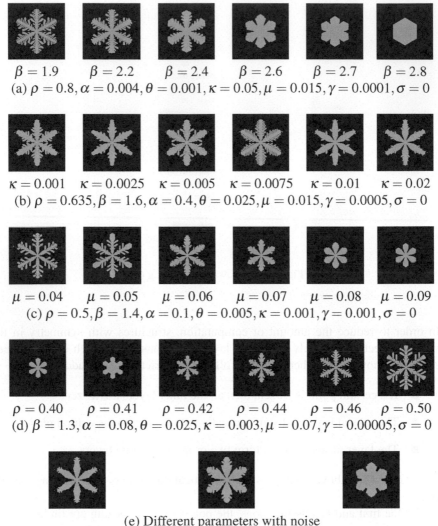

$\beta = 1.9$ $\beta = 2.2$ $\beta = 2.4$ $\beta = 2.6$ $\beta = 2.7$ $\beta = 2.8$
(a) $\rho = 0.8, \alpha = 0.004, \theta = 0.001, \kappa = 0.05, \mu = 0.015, \gamma = 0.0001, \sigma = 0$

$\kappa = 0.001$ $\kappa = 0.0025$ $\kappa = 0.005$ $\kappa = 0.0075$ $\kappa = 0.01$ $\kappa = 0.02$
(b) $\rho = 0.635, \beta = 1.6, \alpha = 0.4, \theta = 0.025, \mu = 0.015, \gamma = 0.0005, \sigma = 0$

$\mu = 0.04$ $\mu = 0.05$ $\mu = 0.06$ $\mu = 0.07$ $\mu = 0.08$ $\mu = 0.09$
(c) $\rho = 0.5, \beta = 1.4, \alpha = 0.1, \theta = 0.005, \kappa = 0.001, \gamma = 0.001, \sigma = 0$

$\rho = 0.40$ $\rho = 0.41$ $\rho = 0.42$ $\rho = 0.44$ $\rho = 0.46$ $\rho = 0.50$
(d) $\beta = 1.3, \alpha = 0.08, \theta = 0.025, \kappa = 0.003, \mu = 0.07, \gamma = 0.00005, \sigma = 0$

(e) Different parameters with noise

Figure 5.9: Snowflake generation.

The update rule of these values is the same as in the literature [41], and the iterative calculation is performed in one step, as follows:

- **Diffusion**: The water vapor amount $d(z)$ at each point of a cell is updated to the weighted average of the neighboring cells' values. The weights are 4 for the six surrounding cells and the same height point of itself, whereas the weights are 3 for the cells above and below itself.

- **Condensation**: For each place in contact with the ice, add $(1 - \kappa) \times d(a)$ to the b as the solidified volume, and subtract the same value from $d(a)$.

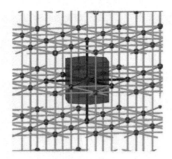

Figure 5.10: Three-dimensional lattice for snowflake growth.

■ **Freeze Determination**: Determines whether each point is frozen or not. Let the number of points frozen at an adjacent point be nt. Then, the point is determined to be frozen if the solidified volume b is equal to or larger than a threshold value. For the threshold, $\beta_{0\,nt}$ is chosen when $a = 0$ (i.e., when there is no frozen layer up and down), and $\beta_{1\,nt}$ is chosen when $a \neq 0$ (i.e., when there is a frozen layer up and down).

■ **Evaporation**: Assuming that some of the water, that was condensed and solidified on the boundary, evaporates, $\mu \times b$ is added to $d(a)$ and the same value is subtracted from b.

Since it is not separated into two or more layers at each point, in the visualization of the three-dimensional structure, the uppermost layer in each cell is expressed as the color intensity. To clarify the factors of crystal formation, the amount of water vapor at $z = 0$ was visualized. Since the amount of water vapor in this surface can be described with $d(0)$, it is visualized in four colors according to this value (see Fig. 5.11). Furthermore, in order to make it easier to understand, we depict the point where crystals are generated in a different color when the frozen amount b at that point exceeds a certain constant value. Since the point where the freezing occurs is only the boundary of the crystal, the point indicated by this point can simultaneously represent the boundary of the crystal.

In the literature [41], water vapor is not supplied from the open air. Therefore, as crystals are formed, the surrounding water vapor is exhausted and the crystal formation is delayed. Moreover, there is a disadvantage that it is impossible to reflect the influence of the change of the humidity of the open air. Therefore, in this simulator, an environment not affected by the crystalline water vapor content was introduced. This environment is called an environmental plane. In the implementation, calculations were made under the assumption that the amount of water vapor on that surface did not change, so that the water vapor was constantly supplied from that side. As a result, water vapor is periodically supplied to the crystal, and it is expected that the generation of crystals can be accelerated. The point where the distance from the center point is the same as the boundary was used with the en-

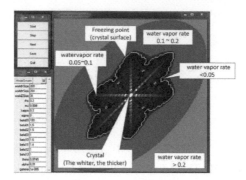

Figure 5.11: Crystal structure for snowflake in different colors.

vironmental plane. When converting the point (x,y) on the lattice to the real-world point (X,Y), the following equation is satisfied:

$$X = x, \quad Y = \frac{2}{\sqrt{3}}y + \frac{1}{\sqrt{3}}x. \tag{5.10}$$

Therefore, when the distance from the center point to the environment plane is r, the point satisfying the equation,

$$4(x^2 + xy + y^2) = 3r^2, \tag{5.11}$$

is set to be the boundary with the environmental surface so that the amount of water vapor at that point is always ρ. In order to avoid the influence of the environment, the simulation is terminated when the crystal grows close to the vicinity of the environment.

Figure 5.11 shows the simulation screen. The part indicated by white in the center is a crystal. The area of light blue surrounding this crystal is being solidified. In addition, the surrounding points indicate that a large amount of water vapor is present in increasing order.

The crystal structures with different values of the initial amount of water vapor ρ are shown in Fig. 5.12. The β value was taken from Case 2 in reference [41], and set $\beta_{10} = 1.6, \beta_{01} = \beta_{02} = 1.5, \beta_{11} = 1.4$, and the others were set to 1.0. Other parameters were fixed with $\kappa = 0.2$ and $\mu = 0.008$. The size of the simulation plane was set to be 300 and 16 layers in height (i.e., $0 \leq z \leq 15$).

From the figure, it can be deduced that the smaller the value of ρ is, the smaller the number of branches becomes. When the ρ value becomes larger, crystals with more complex shapes are generated. In general, it is said that snow crystals tend to have a structure with fewer branches when the amount of water vapor is small, which is consistent with the simulation result.

As we can see the distribution of the amount of water vapor in the simulation, the larger the value of ρ, the larger the amount of water vapor at the tips and

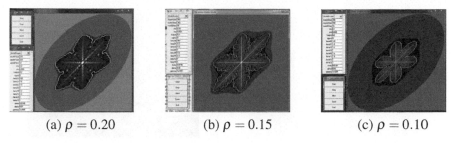

(a) $\rho = 0.20$ (b) $\rho = 0.15$ (c) $\rho = 0.10$

Figure 5.12: Simulation of snowflake growth.

the edges becomes, indicating that the development of the branches is being promoted. On the other hand, when ρ is small, the amount of water vapor between the branches is greatly reduced so that the amount of water vapor at these points decreases and the formation of the branches is delayed. It can be seen that the shape of each branch is determined by this distribution of the water vapor amount.

As described above, by displaying the three-dimensional structure and the amount of surrounding water vapor in real time, it is possible to confirm the generation factor of the crystal structure that was not clear in the conventional studies [41] and observe the formation of the crystals.

5.3   Why do Fish Patterns Change?

5.3.1   Turing Model and Morphogenesis

"Morpho" means "shape", and "genesis" means "generation." Alan Turing believed that the morphogenesis of organisms could be explained by the reaction and diffusion of morphogens, which are hypothetical chemicals, and he proposed the model described below. It is very interesting that von Neumann and Turing, pioneers in computer science, were exploring models of life phenomena in the early days of computers.

Two morphogens, X and Y, activate and inhibit, respectively. X and Y are governed by the following reaction and diffusion equations. x, y are the concentrations of X and Y; f, g are the generation rates of X and Y; D is the diffusion coefficient.

$$\frac{\partial x}{\partial t} = f(x,y) + D_x \nabla^2 x, \tag{5.12}$$

$$\frac{\partial y}{\partial t} = g(x,y) + D_y \nabla^2 y. \tag{5.13}$$

The first term on the right-hand side is a generation term from chemical reaction, and the second term represents movement by diffusion. These are called reaction-diffusion equations.

(a) various shells   (b) chambered nautilus (@PNG in 2005)

Figure 5.13: CA patterns found on shells.

Using this model, Turing hypothesized the following claim:

Stable patterns would form:

- if the evolution of the inhibitor is slower than that of the activator (i.e., $\frac{\partial x}{\partial t} > \frac{\partial y}{\partial t}$),

- and if the diffusion of the inhibitor is faster than the activator (i.e., $D_y y > D_x x$).

In fact, the formation of various patterns can be simulated by changing $f(x,y)$ and $g(x,y)$. The following are simulations of the Turing model using cellular automata (CA).

Kusch et al. defined the CA rules for reaction and diffusion, as shown below, and simulated the Turing model [65]. These are one-dimensional CA to reproduce patterns such as those found on shells and animal fur. Figure 5.13 is a well-known pattern on a shell. The objective here is to generate such patterns.

Each cell has two variables, $u(t)$ and $v(t)$, corresponding to the amount of activator and inhibitor, respectively. $u(t)$ may be 0 or 1, where 0 is the dormant state (white) and 1 is the activated state (black). $u(t)$ and $v(t)$ transition to $u(t+1)$ and $v(t+1)$ through two intermediate steps, each according to the following rules.

```
(1) If v(t)>=1, then v1=[v(t)(1-d)-e],
       else v1=0
(2) If u(t)=0, then u1 = 1 with possibility p,
                 and u1 = 0 with possibility 1-p.
       else u1=1
(3) If u1=1, then v2=v1+w1,
       else v2=v1
```

(4) If u1=0 and nu>{m0+m1*v2}, then u2=1,
    else u2=u1
(5) v(t+1)={<v2>}
(6) If v(t+1)>=w2, then u(t+1)=0,
    else u(t+1)=u2

Here, {} indicates the closest integer, < > is the average within distance rv, and nu is the number of activated cells within distance ru. Statement (1) expresses the decrease in inhibitors per time step. In particular, a linear decrease is observed with $d = 1$ and $e = 1$, and an exponential decrease with $0 < d < 1$ and $e = 0$. Dormant cells are activated at a fixed probability according to (2). Statement (3) shows that activated cells emit inhibitors. Statement (4) states that a cell is activated if the number of activated cells within a distance (nu) is larger than the linear function (m0+m1×v2) of inhibitors (v2). This is the description of diffusion of activators. Statement (5) means that the inhibitor becomes the average within the distance rv, showing how inhibitors diffuse. Statement (6) states that a cell becomes inactive if the inhibitor mass is larger than the constant value.

Figure 5.14 shows the results of experiments along with the parameters of $d = 0.0, e = 1.0$, and initial probability of InitProb$= 0.0$. Other parameters are shown in Table 5.3. Note that branching and interruption, as seen in the figures, are difficult to reproduce in differential equation based models, but are easily reproduced in CA models by using appropriate parameters. Results (e) and (h)∼(k) correspond to Class IV described in section 1.3.

By simulating the Turing model in two dimensions, it is possible to create a variety of repetition patterns by simply changing the parameters. It can be used to realize temporal change of the pattern. For example, the pattern of a fish depends on its age. Figure 5.15 shows the striped pattern of the young and mature Emperor Angelfish. There is also a theory that the patterns, that differ greatly between adults and children, evolved to prevent children from being attacked by their parents because they are regarded as rivals. Animals grow in size by growth, but, assuming that the skin area spreads evenly, the pattern should also grow as it is. On the other hand, if the pattern is a "wave" by the Turing model, pattern reconstruction will occur to keep the interval. Kondo et al. have demonstrated that the branching of the stripes of the pattern on the Emperor Angelfish is the same as the change predicted from Turing's equation [64]. That is, assuming that the pattern is a wave, characteristic points like branches are unstable, so peculiar changes occur. As a result, it was also found out that the number of vertical stripes increases with the number of branches. Furthermore, predictions that the striped pattern moves along the body surface, and the appearance and disappearance of the metastasis in a regular pattern, were proven.

Figure 5.14: Turing model simulation results (see Table 5.3 for the above parameters).

Table 5.3: Parameters for Turing model simulation.

(a)	ru=1,rv=17,w1=1,w2=1,m0=m1=0,p=0.002
(b)	ru=16,rv=0,w1=8,w2=21,m0=0,m1=1,p=0.002
(c)	ru=2,rv=0,w1=10,w2=48,m0=m1=0,p=0.002
(d)	ru=1,rv=16,w1=8,w2=6,m0=m1=0,p=0.002
(e)	ru=1,rv=17,w1=1,w2=1,m0=m1=0,p=0.002
(f)	ru=3,rv=8,w1=2,w2=11,m0=0,m1=0.3,p=0.001
(g)	ru=1,rv=23,w1=4,w2=61,m0=m1=0,p=0,d=0.05,e=0,initProb=0.1
(h)	ru=3,rv=8,w1=2,w2=11,m0=0,m1=0.3,p=0.001
(i)	ru=3,rv=0,w1=5,w2=12,m0=0,m1=0.22,p=0.004,d=0.19,e=0.0
(j)	ru=2,rv=0,w1=6,w2=35,m0=0,m1=0.05,p=0.002,d=0.1,e=0.0
(k)	ru=1,rv=2,w1=5,w2=10,m0=0,m1=0.3,p=0.002

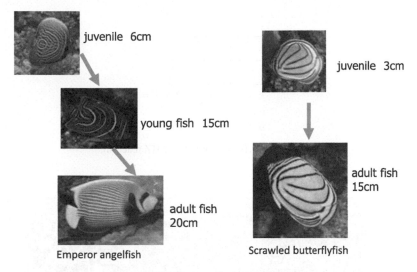

juvenile 6cm

juvenile 3cm

young fish 15cm

adult fish 15cm

adult fish 20cm

Emperor angelfish

Scrawled butterflyfish

Figure 5.15: Pattern changes in fishes.

5.4 BZ Reaction and its Oscillation

In addition to the living body, it is known that waves generated by chemical reactions make rhythmic patterns. The representative is the Belousov-Zhabotinsky reaction (hereinafter referred to as the BZ reaction). This is the reaction that Belousov, a biologist of the former Soviet Union, discovered in the tricarboxylic acid cycle (one of the important reaction systems involved in biological metabolism). The characteristic of this reaction is that the color of the solution changes alternately as a result of the oxidation and reduction repeatedly taking place, thus, changing the concentration of certain substances periodically and spatially. Recently, the BZ reaction has been studied in various ways. An interesting research example is a phenomenon in which a sudden reemergence of vibration occurs after several hours up to 20 hours, even after the vibration has lasted for about 10 hours in a certain concentration region [86].

Let us simulate the BZ reaction using CA. This model can be expressed as follows [28]. Each cell has $n+1$ states of $0, 1, \cdots, n$, and these are classified into one of the three classes, healthy (in case of 0), infected $(1, \cdots, n-1)$, and diseased (n). The state update rule is as follows:

$$n_{t+1} = \begin{cases} \frac{a_t}{k_1} + \frac{b_t}{k_2} & \text{Healthy : When } n_t = 0, \\ \frac{s_t}{a_t + b_t + 1} + g & \text{Infected : When } 1 < n_t < n-1, \\ 0 & \text{Diseased : When } n \leq n_t. \end{cases} \tag{5.14}$$

Here, the state value at time t, denoted as n_t. a_t, is the number of infected adjacent cells, b_t is the number of diseased adjacent cells, s_t is the sum of adjacent cells and

0	1	0
3	1	2
1	0	0

$s = 1 + 1 + 1 + 2 + 3 = 8$

$a = 3$

$b = 1$

Figure 5.16: CA model for BZ reaction.

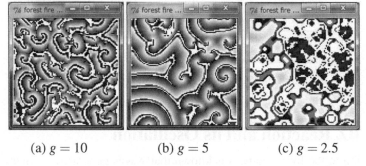

(a) $g = 10$ (b) $g = 5$ (c) $g = 2.5$

Figure 5.17: BZ reaction.

its own state value n_t. Figure 5.16 shows an example of calculation (for $n = 3$) for 8 neighborhoods (Moore neighborhood). Additionally, k_1 and k_2 are constants that determine the progress of the BZ reaction.

The experimental results of the BZ reaction are shown in Fig. 5.17. Here, the initial states of the simulation are set to $k_1 = 3, k_2 = 5, n = 1000$ and g is set to be 10, 5 or 2.5. The greater the g value, the less sick cells (red), and the more infected (gray, the higher state value, the darker) cells are observed. Note that white cells are in healthy state. Furthermore, big g is more likely to result in a steady pattern.

Let us take a look at the time variation of the BZ reaction. Here, parameters are set to be $k_1 = 2, k_2 = 3$ and $n = 200$. The g value is set to be 30 or 70. Two types of neighborhoods are considered: Moore neighborhood and chess knight's movement (i.e., two cells up, one to the right, etc.). Figures 5.18~5.21 show the snapshots of the simulation for each parameter and the transition of the number of sick/healthy cells. The color of the cell is green if it is healthy, red if it is sick, and blue if it is infected (the higher state value, the brighter). Setting $g = 30$ in the Moore neighborhood, the two states in the snapshot in Fig. 5.18 are alternately repeated and vibrate. If $g = 70$ it will gradually settle down from the left state, shown in Fig. 5.19, to the right state. At this time, the vortex pattern occurs and spreads over the whole as the time passes. Looking at the plot of the number of cells, there are

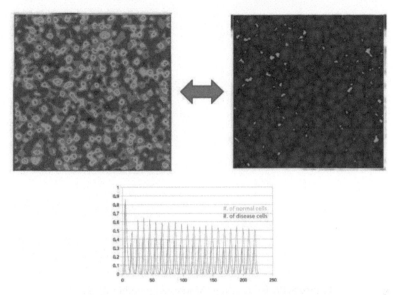

Figure 5.18: BZ reaction (Moore neighborhood), $g = 30$.

Color version at the end of the book

also large waves in addition to fine vibrations, suggesting that this CA has multiple time constants. By using the neighborhood of the knight, when experimenting with $g = 30$, the same vibration phenomenon, similar to the previous one seen in the Moore neighborhood, was observed (Fig. 5.20). On the other hand, if $g = 70$, the same pattern continued without converging (Fig. 5.21).

Additionally, biochemical reactions behave differently when dust or dirt enters the reaction vessel. This process was simulated, as shown in Fig. 5.22. The situation when there is no obstacle is presented in (a). The parameters are set to be $k_1 = 1, k_2 = 1, g = 10$ and $n = 50$. Figure 5.22(b) shows a case where a point obstacle is placed in the upper left. Inserting this single point increased the number of vortex centers by two. In (c) the shape of the reaction vessel is changed, and holes are opened in the middle. At this time, a new vortex is not created, but the diffraction phenomenon can be observed along the boundary of the obstacle.

5.5 Why do We have Mottled Snakes? Theory of Murray

As an interesting application, let us introduce Murray's theory of mammalian coat (fur) pattern [78, 79].

In this theory, the fur pattern of animals seen in nature can be recreated, by changing two numerical parameters. Whether the pattern becomes spotted or striped depends on the timing of the chemical reactions in the skin. Therefore, an-

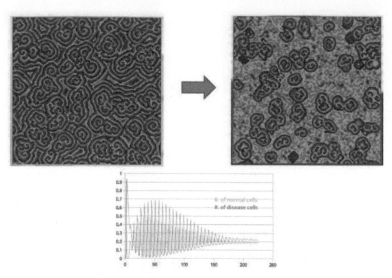

Figure 5.19: BZ reaction (Moore neighborhood), $g = 70$.

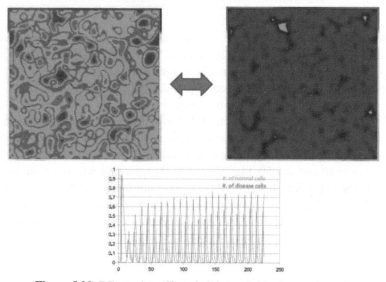

Figure 5.20: BZ reaction (Chess knight's neighborhood), $g = 30$.

imal fur pattern differences can be thought of as a result of purely mathematical rules [114, 27].

If the skin area is very small or large, no pattern will appear. On the other hand, if it is not too wide and too narrow, and the shape is long or thin, stripes appear orthogonal to the major axis of that shape. An example is a zebra. In this animal,

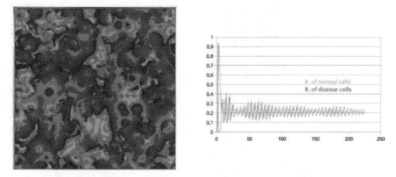

Figure 5.21: BZ reaction (Chess knight's neighborhood), $g = 70$).

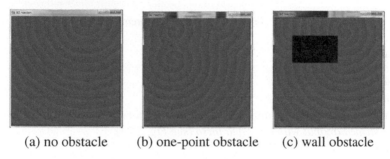

(a) no obstacle   (b) one-point obstacle   (c) wall obstacle

Figure 5.22: BZ reaction (with/without an obstacle).

the fetus has a long pencil-like shape for four weeks in the early pregnancy. Since the chemical reaction occurs at this time, it can be thought to be striped pattern. Spots will appear when the whole shape is close to a square. An example of this is a leopard. In this animal, a spot will be formed as a chemical reaction occurs when the fetus is round. However, the tail is different. During the growth of the fetus, the tail ends up with a striped pattern because it has a long pencil-like shape.

From Murray's theory, the following theorem can be derived as a prediction:

Theorem 5.1
Snake's theorem
Snakes are always striped (rings) and have no spotted patterns.(see Fig. 5.23).

Figure 5.23: Snake's theorem.

Theorem 5.2
Body and tail theorem
There are no spotted tail animals with the striped torso, On the other hand, there is a striped animal with a spotted torso (see Fig. 5.24).

Ironically, the above emergent behavior cannot be explained in evolutionary theory. On the other hand, according to Murray's theory, this can be proved using the following facts:

■ Many animal embryos are composed of a rounded torso and a thin tail.

■ There is no fetus with a round tail and a thin torso.

The smaller the diameter of the tail is, the less likely it is that the striped pattern will become unstable and its instability will be higher for bodies with bigger diameters.

So far, theoretical studies on emergent phenomena of morphogenesis, such as the Turing model and Murray's law, have been explained. One of the most common criticisms about such research is that "It just happened that accidentally chosen parameter recreated the same behavior as real animals." That is, in many studies, one of the sufficient conditions of the pattern formation (the possibility of causing a phenomenon) is defined. On the other hand, what is more important in biology are requirements (this mechanism must work).

These two-way (bidirectional) studies are expected in the future, and, recently, some interesting results have been obtained. For example, cone cells (photoreceptor cells), that correspond to different frequencies of light in the fish retina, form a regular mosaic pattern. On zebrafish retina, there are four patterns of cone cells with

Jaguar

Amur Tiger

Giraffe and zebra

Snow leopard

Figure 5.24: Body and tail theorem.

peaks of sensitivity for the light of blue, red, green, and ultraviolet wavelengths. Mochizuki et al. assumed that the cells would be rearranged based on adhesion, and proposed a CA model to consider the placement of cells in the lattice space [77]. Then, the requirements of intercellular adhesion were mathematically derived for the formation of zebrafish sequences. Furthermore, interactions between cells were predicted, and validation in actual organisms was carried out.

It is also confirmed that there is a chemical reaction system according to the Turing model in reality. This was triggered by the success of the chemical production of reaction-diffusion waves. As an example, the waves moving on the skin of the mouse were observed. This suggests that the Turing wave works not only with fish but also widely as a fundamental principle of skin pattern formation of vertebrates [64].

Chapter 6

Emergence of Intelligence

As one's days dwindle, life begins to shift from an iterative prisoner's dilemma, in which defection can be punished and cooperation rewarded, to a one-shot prisoner's dilemma, If you can convince your children that your soul will live on and watch over their affairs, they are less emboldened to defect while you are alive.

—Steven Pinker

6.1 Evolution of Cooperation and Defection

6.1.1 How to Clean a Fish

There is a fish called Cleaner Wrasse (Labroides dimidiatus). In Japan, it is usually seen in the south sea, and it is a long slender fish of about 10cm in length. This fish is known for cleaning other fish of parasites. When diving in the southern sea, it can be easily observed how these fish enter into the mouth and gills of a large fish, such as the Japanese parrotfish (Fig. 6.1). At this time, it is impressive that a large fish seems to be comfortable opening the mouth wide and letting small Wrasse freely enter and exit. If a large fish closes its mouth, it can easily get food (Wrasse), but such a betrayal never happens. This is an example of co-evolution (symbiosis) between different species. The Parrotfish is cleaned of parasites and leftovers, and Cleaner Wrasse obtains precious food.

However, the story is not so simple. It seems that Wrasse is not always an obedient cleaner. Sometimes it tears off and eats some of the gills during cleaning. In other words, it pinches a little food while cleaning. Certainly, sometimes it can be observed that the fish being cleaned suddenly moves the body as if it trying to scare the Wrasse away. At this time, the Parrotfish punishes the Cleaner Wrasse by chasing it aggressively or swimming away and the Wrasse runs away in a hurry. Normally, however, the Wrasse approaches and begins cleaning again. Punishment seems to lead to a tendency to reduce the betrayal (biting the Parrotfish) of the

Figure 6.1: A big cod fish cleaned by a cleaner fish (@Exmouth, Western Australia in 2014).

Cleaner Wrasse. In fact, based on the field data, the fish requesting to be cleaned punishes by chasing or swimming away from the fish cleaning their body, not the parasite. Experimental results showed, that the Wrasse's tendency to eat parasites increased when it was punished, while the tendency to bite the Parrotfish is decreased [23]. From a recent research [94], it was found out that the Cleaner Wrasse is more prone to betray when there are a few fish around. In other words, if there are eyewitnesses, they act honestly, and when they are not being watched over, they act maliciously. This is a somewhat clever strategy that also considers the effect of attracting customers in the future.

Additionally, to make things more complicated, there are also professional thieves that imitate the Cleaner Wrasse. This false cleanerfish is a different kind of the Blenniidae family. However, the size, mottle, and color are exactly the same as the Cleaner Wrasse. It is said that the position of the mouth is slightly lower than that of the Blenniidae, but it is not possible to distinguish unless very carefully observed. This false cleanerfish has a bad character and pretends to clean the big fish, and after it is allowed to clean the big fish, it runs away after eating off the fin and the skin.

Such fish strategies can be called the betrayal and cooperation. How did that strategy evolve? The appearance of the false cleanerfish that mimics the Cleaner Wrasse occured because there were big fish deceived by it. In other words, the strategy of the false cleanerfish is effective because a fish (Wrasse), that does the cleaning honestly, exists. If a lot of Wrasses abandon honesty and become "thieves," both the false cleanerfish and the Wrasse will be driven away by big fish. On the other hand, from the viewpoint of the big fish, although it is sometimes "robbed," it should be better to accept the "sweepers," which work honestly in most cases even if they have to make a little sacrifice. This is because the false cleanerfish and the Wrasse cannot be distinguished from the outside. However, if the number of Wrasse is reduced, and only the false cleanerfish stay, the story will be different. The symbiotic relationship between Wrasse, false cleanerfish and the big fish to be cleaned is founded on a delicate balance. Wrasse's betrayal ("pinching food") may be a remnant of the evolutionary process or a new parasitic relationship.

The relationship between betrayal and cooperation, as described above, is modeled in the framework of a game theory called the Prisoner's Dilemma.

6.1.2   The Prisoner's Dilemma

Two suspects (A and B) have been arrested by the police for a certain crime (Fig. 6.2). Although the suspects are accomplices and their guilt is strongly suspected, insufficient evidence exists to prove their guilt beyond doubt, and, therefore, the police separate the suspects and wait for a confession. A and B can adopt one of two strategies, namely, to confess (i.e., to defect, denoted by D below) or refuse to confess (i.e., to cooperate, denoted by C below), as shown in Table 6.1. The case where A does not confess is labeled A1, and the case where A confesses is labeled A2. Similarly for B, B1 denotes a refusal to confess and B2 denotes a confession. If neither suspect confesses, both of them would receive two-year sentences; however, if only one of them confesses, the one confessing would be sentenced to only one year due to extenuating circumstances, while the other suspect would receive a five-year sentence. Finally, if both suspects confess, both of them would receive three-year sentences.

Therefore, although the sentence is only two years for both suspects if neither of them confesses, the loss for one suspect is considerably greater if a confession is obtained from only their accomplice. The dilemma here is whether A and B should confess. In terms of pure strategy, it is straightforward to show that the pair (A2, B2) is an equilibrium point, and that no other equilibrium points exist. Here, the equilibrium point indicates that if either suspect adopts confession as their strategy, their accomplice is at a disadvantage if they do not adopt the same strategy[1]. Conversely, if the suspects adopt the strategic pair (A1, B1), neither suspect confesses in the belief that their accomplice will not confess either, and the sentence for both of them is only two years. This sentence is obviously a more favorable outcome than the equilibrium point for both suspects, but since the suspects are separated from each other when taken into custody, they are unable to cooperate. This type of game is known as the prisoner's dilemma.

Situations such as those arising in the prisoner's dilemma are also often encountered in biology as well as in human society. For example, if an appropriate incentive is present, the following situations can be modeled similarly to the prisoner's dilemma.

■ Social interaction between animals in nature:

   ■ Social grooming in primates.
   ■ Cleaner fish and the fish that they clean.
   ■ Parasites and hosts.

---

[1] Detailed definition and discussion are given later in page 164.

• Not confession   Cooperate
• Confession   Defect

Figure 6.2: The prisoner's dilemma.

Table 6.1: Benefit and cost in the prisoner's dilemma.

(a) Benefit and cost from the viewpoint of A

	B_1 B: not confess (C)	B_2 B: confess (D)
A_1 A: not confess (C)	−2 Two years of imprisonment	−5 Five years of imprisonment
A_2 A: confess (D)	−1 One year of imprisonment	−3 Three years of imprisonment

(b) Benefit and cost from the viewpoint of B

	B_1 B: not confess (C)	B_2 B: confess (D)
A_1 A: not confess (C)	−2 Two years of imprisonment	−1 One year of imprisonment
A_2 A: confess (D)	−5 Five years of imprisonment	−3 Three years of imprisonment

- Social interaction between humans:

    - Interaction between countries.
    - Interaction between tribes.

Payoff is considered using a game theory concept of "equilibrium point." The equilibrium point is defined as follows:

Definition 6.1   Equilibrium point
Equilibrium point is a strategy characterized by the notion that if an opponent chooses a strategy, one must also choose the same strategy to avoid a loss.

More precisely, consider the following payoff table related to two players:

$$
\begin{array}{cc}
& X \quad Y \\
\begin{array}{c} X \\ Y \end{array} & \begin{pmatrix} a & b \\ c & d \end{pmatrix}
\end{array}
\tag{6.1}
$$

In other words,

■ In the match $X \times X$, both gain a.

■ In the match $Y \times Y$, both gain d.

■ In the match $X \times Y$, X gains b and Y gains c.

In this case, we have:

■ If $a > c$, then strategy X is a strict Nash equilibrium point.

■ If $a \geq c$, then strategy X is a Nash equilibrium point.

■ If $d > b$, then strategy Y is a strict Nash equilibrium point.

■ If $d \geq b$, then strategy Y is a Nash equilibrium point.

This idea is attributed to John Nash[2]. Using Brower's fixed-point theorem, he proved that a non-collaborative game with at least three players will always have an equilibrium point.

In the prisoner's dilemma game, it is easy to verify that a pure combination of strategies (A2, B2) represents an equilibrium point. Moreover, no other equilibrium point exists. However, if one does not confess, believing that the other will not do it either, characterizing an (A1, B1) combination, both will be sentenced to only 2 years. This result is better than the equilibrium points for both; however, since each prisoner is in solitary confinement and cannot cooperate, the players fall into a dilemma. Point (A1, B1) is "Pareto efficient," where a strategy change would bring an advantage to a player, but he refrains from doing it if that results in a disadvantage to the other one; this is a demonstration of care. In other words, this is a combination of strategies that respects the common part of both players' preferences. In the prisoner's dilemma, a dilemma is characterized by the fact that the Nash equilibrium point is not Pareto efficient.

---

[2]John Nash (1928–2015): American economist and mathematician. In 1994, he received the Nobel Prize in economics. The American movie "Beautiful Mind (2001)" is a mystery story that depicted a part of his life and won many Academy Awards. An interesting quote from the movie is that what Adam Smith wanted to say can be mathematically proved now.

6.1.3 Iterated Prisoner's Dilemma

This section considers an extended version of the prisoner's dilemma, known as the "iterated prisoner's dilemma" (IPD), in which the prisoner's dilemma is repeated a number of times with the same participants, and a final score is obtained as the sum of the scores in all iterations. Hereinafter, the choice made at each step is referred to as a move, or, more precisely, a move indicates either cooperation (C: no confession) or defection (D: confession).

As with the prisoner's dilemma, situations similar to that in IPD can regularly be observed in nature and in human society. A well-known example is regurgitation in vampire bats [24]. These small bats, living in Central and South America, feed on mammalian blood at night. However, their bloodsucking endeavors are not always successful, and at times they face starvation. Therefore, vampire bats that feed successfully will regurgitate part of their food and share it with other vampire bats that were unable to find food. Bats who receive food in this manner later return the favor. Wilkinson et al. observed 110 instances of regurgitation in vampire bats, of which 77 cases were from a mother to her children, and other genetic relations were typically involved in the remaining cases. Nevertheless, in several cases, regurgitation was also witnessed between companions sharing the same den, with no genetic affinity. Modeling this behavior on the basis of the prisoner's dilemma, where cooperation is defined as regurgitation and defection is defined as the lack thereof, yields results similar to those in Table 6.2. A correlation between weight loss and the possibility of death due to starvation has also been found in bats. Therefore, the same amount of blood is of completely different value to a well-fed bat immediately after feeding and to a bat that has almost starved to death. In addition, it appears that individual bats can identify each other to a certain extent, and, as a result, they can conceivably determine how a certain companion has behaved in the past. Thus, it can be said that bats will donate blood to "old friends."

To set up the IPD, one player is denoted as P_1 and the other party is denoted as P_2. Table 6.1 is then used to create Table 6.3, describing the benefit for P_1. In this table, larger values indicate a higher benefit (which, in the prisoner's dilemma, is in the form of shorter imprisonment terms for the suspects).

Here, the following two inequalities define the conditions required for the emergence of the dilemma:

$$T > R > P > S, \tag{6.2}$$

$$R > \frac{T+S}{2}. \tag{6.3}$$

The first inequality is self-explanatory, while the implications of the second one will be explained later.

As a possible strategy in IPD (iterated prisoner's dilemma: Iterated PD), for example, the following ones can be thought of:

Table 6.2: The strategy of vampire bats [24].

	Companion cooperates.	Companion defects.
I cooperate.	**Reward: high** On the night of an unsuccessful hunt, I receive blood from you and avoided death by starvation. On the night of a successful hunt, I donate blood to you, which is an insignificant expenditure to me at that time.	**Compensation: zero** Although on a successful night I donate blood to you and save your life, you do not donate blood to me when my hunt is unsuccessful and I face death by starvation.
I defect.	**Temptation: high** You save my life when my hunt is unsuccessful. However, on a successful night, I do not donate blood to you.	**Penalty: severe** I do not donate blood to you even when my hunt is successful, and death through starvation is a real threat to me on the night of an unsuccessful hunt.

Table 6.3: Codes and benefit chart in IPD.

	The other party (P_2) does not confess. $P_2 = C$	The other party (P_2) confesses. $P_2 = D$
I (P_1) do not confess. $P_1 = C$	Benefit: 3 Code: R	Benefit: 0 Code: S
I (P_1) confess. $P_1 = D$	Benefit: 5 Code: T	Benefit: 1 Code: P

1. All-C: Always cooperate regardless of the opponent's hand (do not confess).

2. All-D: Always betray regardless of the opponent's hand (confess).

3. Repeat cooperation and betrayal.

4. RANDOM: Decide a hand for each round at random.

5. GRIM: C for the first time, C until the opponent betrays, and forever D after betrayal, and never forgives.

6. GRIM*: It is the same as GRIM, but at the end it must be D ("parting shot").

7. WSLS: Win and Lose (Win-Stay-Lose-Shift) strategy, also called Pavlov strategy. See Table 6.5.

The cooperative solution, which does not appear in a single game, evolves in IPD. Cooperation becomes inevitable since defection entails subsequent revenge. An example of a well-known cooperation strategy is "tit for tat" (TFT):

TFT (tit for tat)

1. The first move is performed at random.

2. All subsequent moves mirror P_2's move in the previous iteration.

In other words, this strategy follows the maxim "Do unto others as they have done unto you (an eye for an eye)". Thus, P_1 confesses if P_2 has confessed in their previous move, and vice versa. This strategy is recognized as being strong in comparison with other strategies (i.e., the total score is usually high). Variants of TFT include the following.

1. Tolerant TFT: Cooperate until P_2 defects in 33% of their moves.

2. TF2T: Defect only if P_2 has defected twice in a row.

3. Anti-TFT: The opposite strategy to TFT.

The benefit at each move and the total score are compared for All-C, All-D, TFT, and Anti-TFT strategies when pitted against each other in Table 6.4. The following can be inferred from the results presented in the table:

1. TFT is equivalent or superior to All-C.

2. TFT is generally superior to Anti-TFT. However, Anti-TFT is superior if P_2 adopts the All-C strategy.

3. All-C yields a lower score against Anti-TFT and All-D; however, it performs well against a TFT strategy.

4. All-D can outperform TFT.

In 1979, Axelrod[3] extended an invitation to a number of game theorists and psychologists to submit their strategies for IPD. Subsequently, a tournament was held using the 13 submitted strategies in addition to the RANDOM strategy. In this tournament, each strategy was set against the others in turn, including itself, in a round-robin pattern. To avoid stochastic effects, five matches were held for each strategic combination. Each match consisted of 200 iterations of the prisoner's dilemma in

---

[3]Robert Axelrod(1943–): American political scientist.

Table 6.4: Comparison of the performance of the four strategies when pitted against each other.

Strategy of P_1	Strategy of P_2 (score / total score)			
	All-C	TFT	anti-TFT	All-D
All-C	3333/12	3333/12	0000/0	0000/0
TFT	3333/12	3333/12	0153/9	0111/3
anti-TFT	5555/20	5103/9	1313/8	1000/1
All-D	5555/20	5111/8	1555/16	1111/4

order to ensure a sufficiently long interaction, as described below. Rather than considering the outcome of individual matches, the ultimate ranking was determined on the basis of the total score.

The winner of the contest was the TFT strategy proposed by Anatol Rapoport[4], a psychologist (and philosopher) at the University of Toronto. This program was the smallest of those submitted, consisting of only four lines of BASIC code. Upon analyzing the results, Axelrod derived the following two morals:

1. One should not deliberately defect from the other party (politeness).

2. Even if the other party defects, one should respond with defection only once and should not hold a grudge (tolerance).

Any strategy abiding by these morals is referred to as being "nice". Axelrod considered that a superior strategy could be derived based on these two morals. He held a second tournament, this time running a wide advertising campaign asking for entries through general computer magazines and publicizing the above two morals as a reference for the participants. Although this call for entries attracted a broad spectrum of participants from six countries, the TFT strategy proposed by Rapoport again emerged victorious, and Axelrod stated various other morals upon analyzing the results from both contests [7].

Looking at the scores obtained in the tournaments, nice strategies similar to TFT dominated the top half of the list. In contrast, the majority of the strategies in the bottom half of the list involved deliberate defection, while the lowest score was obtained by the RANDOM strategy. The question naturally arises as to why this set of circumstance occurs.

Let us consider the strength of the TFT strategy. First, we focus on the case where two nice strategies are competing. Since both strategies denounce deliberate defection, each of the 200 moves is one of cooperation. As a result, the two parties receive a total of $3 \times 200 = 600$ points. This situation is different when a nice strategy is competing against one that promotes defection. For example, consider

---

[4]Anatol Rapoport (1911–): Russian-born Jewish American mathematical psychologist. His vast contribution includes experiments on game theory and related research on social psychology, as well as peace studies.

competition between JOSS and TFT strategies. JOSS behaves almost identically to TFT; however, it defects on random occasions, corresponding to a player with a nice strategy occasionally being tempted to cheat. Thus, JOSS occasionally obtains five points by outperforming TFT (which supports cooperation and denounces defection), and although they are almost identical, JOSS appears to provide a greater benefit.

In the competition, TFT and JOSS cooperated for the first five moves.

```
TFT:   CCCCC
JOSS:  CCCCC
```

At this point, JOSS switched its tactics and defected (cheated).

```
TFT:   CCCCCC
JOSS:  CCCCCD
            ^
```

Although at the next move JOSS returned to its nice tactics, TFT took revenge by defecting.

```
TFT:   CCCCCCD
JOSS:  CCCCCDC
            ^
```

From here on, the moves of the two strategies alternated between defection and cooperation,

```
TFT:   CCCCCCDCDCD
JOSS:  CCCCCDCDCDC
            ^^^^^
```

and the respective scores at that juncture became

```
JOSS:  5 0 5 0 c
TFT:   0 5 0 5 c
```

giving an average of $\frac{0+5}{2} = 2.5$, which is less than the 3 points obtained if the two parties had continually cooperated (cf. the assumption in eq. (6.3)).

Next, JOSS cheated again and defected on the 25th move,

```
TFT:   CCCCCCDCDCD...D
JOSS:  CCCCCDCDCDC...D
            ^
```

resulting in reciprocal defection until the end of the match.

```
TFT:    CCCCCCDCDCD...DDDDDDD
JOSS:   CCCCCDCDCDC...DDDDDDD
                  ~~~~~~~
```

This chain of enmity yielded an average score of 1, and the match ended unchanged since neither side was able to forgive the other's defection and restore cooperation.

At the conclusion of the match, the total score was 241 to JOSS and 236 to TFT, much lower than the 600 points that they could have obtained if they had cooperated throughout.

Hence, TFT has the following characteristics:

1. TFT never attacks the other party. The score obtained by TFT is the same or lower than that of the other party.

2. TFT does not hold a grudge. TFT seeks revenge only once, rather than punishing the other party repeatedly. In this way, TFT does not disregard the chance for reconciliation with the other party.

3. TFT does not require a victim, in other words, it does not exploit the other party. As a result, two players adopting the TFT strategy can enjoy a mutually beneficial relationship.

Conversely, although the All-D strategy can obtain the highest score of 5 points when competing against the All-C (victim) approach, the performance of using All-D is lower when competing against a different strategy.

An important assumption in IPD is that the players do not know when the competition ends, otherwise it would allow for tactics based on "the ultimate defense measure" (always defecting on the last move). In addition, cooperative evolution requires a long competition since a relationship of trust, which is a prerequisite for cooperation, cannot be established in the short term. This idea can be explained using examples such as a student canteen (which adopts an honest business policy of expecting repeat clients in order to establish a long-standing relationship with students) and seaside noodle shops (which survive due to first-time customers, even with expensive and unappetizing food).

Some disadvantages of TFT include

■ In response to noise (when unintentional actions are made by mistake), TFT's moves begin to alternate between cooperation and defection.

■ TFT cannot exploit the All-C strategy (since it usually cooperates).

To address these issues in TFT tactics, the so-called Pavlov strategy (or WSLS, Win-Stay-Lose-Shift strategy) has been proposed, whereby P_1's current move is determined not only by the previous move of P_2, but also by P_1's preceding move

Table 6.5: Comparison between Pavlov and TFT strategies.

preceding move P_1 P_2	Pavlov strategy	TFT strategy
C C	C	C
C D	D	D
D C	D	C
D D	C	D

(Table 6.5). For example, if an incorrect choice has been made (e.g., if P_1 cooperates or defects, expecting P_2 to cooperate, and instead the P_2 defects), the previous choice is reversed in the current move. On the contrary, if P_2 cooperated while P_1 defected in the previous move, changing this choice in the current move is not necessary. The Pavlov strategy is considered to be robust with respect to noise. For example, in a competition where both players have adopted the Pavlov strategy, even if the cooperative relationship is broken due to noise and results in moves (C,D), a cooperative relationship is returned to via moves (D,D) followed by (C,C). Nevertheless, this does not imply that the Pavlov strategy is superior to the TFT strategy. For instance, although the TFT strategy consistently defects against the All-D strategy, the Pavlov strategy alternates between cooperation and defection, and its average score is lower than that for TFT.

IPD has been extended in various ways and has been actively used in research in the fields of AI, AL and evolutionary economics [60]. Examples include studies in which strategic sexual selection is implemented by introducing the concept of gender, and studies where competition is extended to include three or more individuals, resulting in the emergence of cooperation and conspiracy.

Axelrod added norms to the prisoner's dilemma [8]. Here, defectors may be socially sanctioned by reduction of their score. Defecting players had a possibility to be detected by other players, and a strategy to decide whether to penalize a defector that was found (whether to reduce the score of the defector) was added to the strategy to play the prisoner's dilemma.

In other words, when a betrayal action is found, social sanctions against it are inflicted in the form of a penalty. When a player betrays, the possibility that other players will witness it is incorporated. In addition to the strategy of playing the Prisoner's Dilemma, a strategy to decide whether to punish the traitor when witnessed (whether to deduct the traitor's score or not) is integrated.

In each generation, each player performs a single game with all the other players. Every time the player cheats during the match, that action may be witnessed by other players. Each time it is witnessed, the traitor is punished according to the probability set as the strength of the witness's vengefulness. The evolutionary process is applied after all the matches are over. At this time, it is probable that the child's strategy mutates. Sometimes children can obtain boldness and the strength of vengefulness disparate from their parents.

If the social norms are not set at all, the group will eventually overflow with traitors. Since cooperative relationships are not necessarily evolved only with norms, additional meta-norms are also introduced. This is the norm to punish those who pretend to be blind. In other words, when a player who punishes players who pretend not to see traitors is introduced, the latter evolves to punish the traitors and the player who was punished for betrayal was seen to tend to evolve to cooperate.

6.1.4 ESS: Evolutionarily Stable Strategy

Evolutionarily Stable Strategy (ESS) corresponds to a situation when a strategy occupies a group and prevents other strategies from invading it. This idea was advocated in 1973 by John Maynard Smith[5] and George Price[6]. When the strategies A and B are given, and the gain for A is assumed to be $E(A, B)$, then the conditions of the ESS are as follows:

ESS: Evolutionarily Stable State
Strategy I is ESS if either one of the following two conditions is satisfied for any strategy J ($\neq I$):

■ $E(I, I) > E(J, I)$ or

■ $E(I, I) = E(J, I)$ and $E(I, J) > E(J, J)$.

If $E(I, I) = E(J, I)$ and $E(I, J) > E(J, J)$ holds true, this is known as "weak ESS."

As mentioned above, no other strategies can invade a group with the ESS strategy. Narrow Nash equilibrium has the same characteristics, but it does not hold for Nash equilibrium.

All-C is obviously not an ESS. For example, if All-D invades a group of only All-C, the All-C group collapses with All-D being the sole winner. Interestingly, it is known that the TFT is not an ESS.

The following relationship between the solution of Nash equilibrium and ESS strategy holds true:

Narrow Nash equilibrium $\Longrightarrow$ ESS $\Longrightarrow$ Weak ESS $\Longrightarrow$ Nash equilibrium

[5] John Maynard Smith (1920-2004): British biologist. Received the Kyoto Prize in 2001. He is said to be a biologist who had the most influence on biology in the twentieth century.

[6] George R. Price (1922-1975): American population geneticist. Turned from atheism to Christianity, and gave all his wealth to the poor and becomes homeless as result. A book "The Price of Altruism" depicting the sublime way of life leading up to suicide is worth reading [44].

Let us consider the GRIM strategy next. This is C for the first time, C until the opponent betrays, and forever D after it is betrayed and will never forgive. This strategy is regarded as a symbol of the reciprocal behavior of human beings [23]. One example of this is the diamond market in Antwerp (Belgium). Here, transactions are performed in a relatively small group of experts. The trader is given a diamond in the bag to check the item at home. If a person steals it, he will be permanently expelled from the community. The GRIM strategy never tolerates a betrayal, but otherwise places full trust.

The payoff matrix of the m time matchup between All-D and GRIM is as follows [85]:

$$
\begin{array}{cc}
 & \text{GRIM} \qquad\qquad \text{ALL-D} \\
\begin{array}{c} \text{GRIM} \\ \text{ALL-D} \end{array} &
\left(\begin{array}{cc} mR & S+(m-1)\cdot P \\ T+(m-1)\cdot P & mP \end{array} \right).
\end{array} \tag{6.4}
$$

Therefore, if $mR > T + (m-1)\cdot P$ then GRIM becomes a Nash equilibrium in a narrow sense against All-D. All-D cannot invade when all the members of the group are GRIM. However, All-D also becomes a Nash equilibrium point in a narrow sense, since the $mP > S + (m-1)\cdot P$.

Let us look at a matchup between GRIM* (same as GRIM, but is always D at the end) and GRIM. The payoff matrix of the m time matchup is as follows:

$$
\begin{array}{cc}
 & \text{GRIM} \qquad\qquad \text{GRIM*} \\
\begin{array}{c} \text{GRIM} \\ \text{GRIM*} \end{array} &
\left(\begin{array}{cc} mR & S+(m-1)\cdot R \\ T+(m-1)\cdot R & P+(m-1)\cdot P \end{array} \right).
\end{array} \tag{6.5}
$$

It is clear that GRIM* is more advantageous than GRIM, and GRIM* evolutionarily invades the population made up of GRIM.

Now, once everyone becomes GRIM*, they will betray each other even one round before the last round. If betrayal happens one round before the final round, the second from the last round will also end up with everyone betraying each other. That is, GRIM* is more advantageous than GRIM, and GRIM** is more advantageous than GRIM*. This argument will be continued indefinitely, and in the end, All-D will be advantageous. In other words, it was shown that All-D is the solution of the Nash equilibrium in a narrow sense and the only ESS.

Consider the strength of the TFT again. In the battle between TFT and All-D, the payoff matrix becomes the same as GRIM vs All-D payoff matrix, as is shown below:

$$
\begin{array}{cc}
 & \text{TFT} \qquad\qquad \text{ALL-D} \\
\begin{array}{c} \text{TFT} \\ \text{ALL-D} \end{array} &
\left(\begin{array}{cc} mR & S+(m-1)\cdot P \\ T+(m-1)\cdot P & mP \end{array} \right).
\end{array} \tag{6.6}
$$

As it was previously mentioned, when m exceeds the threshold of $\frac{T-P}{R-P}$, i.e., $m > \frac{T-P}{R-P}$, then GRIM was stable against invasion by All-D. On the other hand, TFT is advantageous in comparison with GRIM, in that it starts cooperating again if the other party cooperates. Unlike GRIM, TFT can escape from the world of eternal betrayal.

A few interesting findings have resulted from recent research on IPD[95]. First, contrary to intuition, in IPD, the amount of past memory (history) when playing games over and over does not produce a statistically significant effect. In other words, players with a long memory and players with a very short one (who define strategies based solely on the immediately previous game) tend to achieve the same score when IPD is carried out for an extremely long time.

Considering the above, a strategy called zero-determinant (ZD) was devised for players who memorize just one game backwards. That was an optimal strategy derived in terms of linear algebra. This strategy allows a player to unilaterally set a linear relationship between the player's own payoff and the co-player's payoff regardless of the strategy of the co-player [55]. However, it only works under specific conditions and does not always exist[7]. It is shown that, when a ZD strategy is employed, even if the other player changes strategies in an evolutionary manner (adaptively) the changes end up being fruitless.

6.1.5 IPD using GA

This section describes an experiment based on research conducted by Cohen et al. [20], where an IPD strategy evolves by using a GA. As before, the first player is denoted as P_1 and the other party is denoted as P_2, and the benefit chart in Table 6.3 is used to score the moves.

In this experiment, errors (i.e., when an unintentional move is made) and noise (where the move communicated to the opponent is different from the actual move) are introduced, and the emergence of cooperation in the case of incomplete information is examined. There are 256 players in the IPD, arranged on a 16×16 lattice, and the competition between players consists of four iterations. A single experiment consists of 2,500 periods, where a single period comprises the following steps:

■ Competitions between all players.

■ Learning and evolution of strategies in accordance with an appropriate process.

The basic elements of the experiment are as follows.

---

[7]It is necessary to hypothesize that the player's expected payoff is generated in the steady state of a Markov process.

1. Strategy space–an expression of the strategy adopted by a player, which is divided into the following two types:

   (a) Binary: Pure strategy.

   (b) Continuous: Mixed strategy (stochastic strategy).

2. Interaction process—the set of opponents competing with a player is referred to as the neighborhood of that player. The neighborhood can be chosen in the following six ways:

   (a) 2DK (2 Dimensions, Keeping): The neighbors above, left, below, and right (NEWS: North, East, West and South) of each player on the lattice are not changed, and the neighborhood is symmetrical. In other words, if B is in the neighborhood of A, then A is necessarily in the neighborhood of B.

   (b) FRNE (Fixed Random Network, Equal): For each agent, there is a neighborhood containing a fixed number of agents (four). These agents are selected at random from the entire population and remain fixed until the end. The neighborhood is symmetrical.

   (c) FRN (Fixed Random Network): Almost the same as FRNE, although with a different number of agents in the neighborhood of each agent. The neighborhood is unsymmetrical. The average number of agents is set to 8.

   (d) Tag: A real number between 0 and 1 (a tag or sociability label) is assigned to each agent, introducing a bias whereby agents with similar tags can more easily compete with each other.

   (e) 2DS (2 Dimensions, 4 NEWS): The same as 2DK, except that the agents change their location at each period, thus, NEWS is set anew.

   (f) RWR (Random-With-Replacement): A new neighborhood can be chosen at each period. The average number of agents in the neighborhood is set to 8, and the neighborhood is unsymmetrical.

3. Adaptation Process: The evolution of a strategy can follow any of the following three paths:

   (a) Imitation: Searching the neighborhood for the agent with the highest average score and copying its strategy if this score is larger than one's own score.

   (b) BMGA (Best-Met-GA): Almost the same as Imitation; however, an error is generated in the process of copying. The error can be any of the following.

      i. Error in comparing the performance: The comparison includes an error of 10%, and the lower estimate is adopted.

    ii. Miscopy (sudden mutation): The mutation rate for genes in the mixed strategy (Continuous) is set to 0.1, with the addition of Gaussian noise (mean, 0; standard deviation, 0.4). In addition, the rate of sudden mutation per gene in the pure strategy (Binary), in which probabilities p and q are flipped (i.e., 0 and 1 are swapped), is set to 0.0399. When p mutates, probability i is also changed to the new value of p.

  (c) 1FGA (1-Fixed-GA): BMGA is applied to agents selected at random from the entire population.

The strategy space is expressed through the triplet (i, p, q), where i is the probability of making move C at the initial step, p is the probability of making move C when the move of the opponent at the previous step was C, and q is the probability of making move C when the move of the opponent at the previous step was D.

By using this representation, various strategies can be expressed as follows:

$$
\begin{array}{ll}
\text{All-C} & i = p = 1, \ q = 1 \\
\text{All-D} & i = p = 0, \ q = 0 \\
\text{TFT} & i = p = 1, \ q = 0 \\
\text{anti-TFT} & i = p = 0, \ q = 1
\end{array}
$$

Here, the value of each of i, p and q is either 0 or 1. This form of representation of the strategy space is known as a binary (or pure) strategy.

Conversely, the values of p and q can also be real numbers between 0 and 1, and this representation is known as a continuous strategy. In particular, since 256 players took part in the experiment conducted by Cohen, the initial population was selected from among the following combinations:

$$
p \ = \ \frac{1}{32}, \frac{3}{32}, \ldots, \frac{31}{32}, \tag{6.7}
$$

$$
q \ = \ \frac{1}{32}, \frac{3}{32}, \ldots, \frac{31}{32}. \tag{6.8}
$$

Specifically, each of the 16×16 combinations is set as a single agent (player) with values p, q. Moreover, the value of i is set to that of p.

The performance (the average score of all players in each competition) is calculated as the average value over 30 runs. Note that the occurrence of defection is high if the average value of the population performance is close to 1.0, while cooperation dominates if the value is close to 3.0. Nevertheless, the value does not necessarily reach 3.0 even if the entire population adopts the TFT strategy due to (1) sensor noise in the comparison and (2) copying errors in the adaptation process.

The score of the population can be obtained from the following measures:

■ Av1000: The average score of the last 1000 periods.

■ NumCC: The number of periods where the average score for the entire population was above 2.3.

■ GtoH: Out of 30 iterations, the number of iterations in which the average score for the entire population was above 2.3.

■ FracH: The proportion of periods with a score exceeding 2.3 after the average score for the entire population has reached 2.3.

■ FracL: The proportion of periods with a score below 1.7 after the average score for the entire population has reached 2.3.

■ Rstd: The averaged score distribution for each period during iteration.

Here, an average score of 2.3 is regarded as the threshold above which the population is cooperative, and a score below 1.7 indicates a population is dominated by defection.

The results for all interaction and adaptation processes, in addition to all combinations of representations in the experiments ($2 \times 6 \times 3 = 36$ combinations), are shown in Tables 6.6 and 6.7.

In conducting the experiments, an oscillatory behavior at the cooperation level was observed for all combinations, namely, the average score oscillated between high (above 2.3) and low (below 1.7) cooperation. This oscillation can be explained from repetition of the following steps.

1. If a TFT player happens to be surrounded by TFT players, defecting players are excluded.

2. TFT players soon establish a relationship of trust with other TFT players, and, over the course of time, the society as a whole becomes considerably more cooperative.

3. Defecting players begin to exploit TFT players.

4. Defecting players take a course to self-destruction after exterminating the exploited players.

5. Return to 1.

Two important factors exist in the evolution of cooperation between players. The first factor is preservation of the neighborhood in the interaction process, where the neighborhood is not limited to two-dimensional lattices. A cooperative relationship similarly emerges in fixed networks, such as FRN. In addition, agents that

Table 6.6: IPD benefit chart and codes (Binary representation).

Interaction process	Adaptation process	Av1000	NumCC	GtoH	FracH	FracL	Rstd
2DK	1FGA	2.560 (.013)	30	19	.914	.001	0.173 (.017)
	BMGAS	2.552 (.006)	30	10	.970	.000	0.122 (.006)
	ImitS	3.000 (.000)	30	7	1.00	.000	0.000 (.000)
FRNE	1FGA	2.558 (.015)	30	21	.915	.000	0.171 (.012)
	BMGAS	2.564 (.007)	30	9	.968	.000	0.127 (.007)
	ImitS	2.991 (.022)	30	6	1.00	.000	0.000 (.000)
FRN	1FGA	2.691 (.008)	30	22	.990	.000	0.120 (.010)
	BMGAS	2.629 (.010)	30	14	.913	.006	0.229 (.016)
	ImitS	1.869 (1.01)	13	7	1.00	.000	0.000 (.000)
Tag	1FGA	2.652 (.010)	30	15	.975	.000	0.132 (.010)
	BGMAS	1.449 (.186)	30	255	.191	.763	0.554 (.170)
	ImitS	1.133 (.507)	2	6	1.00	.000	0.000 (.000)
2DS	1FGA	2.522 (.024)	30	26	.867	.000	0.197 (.020)
	BMGAS	2.053 (.128)	30	95	.443	.280	0.532 (.064)
	ImitS	1.000 (.000)	0	-	-	-	0.000 (.000)
RWR	1FGA	2.685 (.0009)	30	54	.985	.000	0.127 (.013)
	BMGAS	1.175 (.099)	13	972	.191	.763	0.109 (.182)
	ImitS	1.000 (.000)	0	-	-	-	0.000 (.000)

stochastically select their neighborhoods by using tags are more cooperative than agents with random neighborhoods.

The second factor is the process of generating and maintaining the versatility of the strategy. For example, cooperation is not common in adaptation processes that are similar to Imitation. Since errors do not exist in Imitation, the versatility of the strategy is soon lost, which often results in an uncooperative population. However, on rare occasions, extremely cooperative populations arise even in Imitation, and Imitation was the strategy producing the highest scoring population.

On the other hand, it is known that cooperative behavior may be difficult to evolve depending on the structure of the network in the neighborhood. In particular, recent studies [119] have shown that cooperative behaviors are easier to evolve in heterogeneous networks than in homogeneous networks[8]. There is also a report that, when humans play IPD games, the cooperation evolves in a similar manner in both the homogeneous and the heterogeneous networks [38].

6.1.6 IPD as Spatial Games

When the prisoner's dilemma is configured as a spatial game, it results in interesting patterns, such as the "big bang" or the "kaleidoscope" [85].

---

[8]Networks with a regular lattice-like structure of the neighborhood are called homogeneous networks. On the other hand, if the network has a randomness or scale-free structure, it is heterogeneous.

Table 6.7: IPD benefit chart and codes (Continuous representation).

Interaction process	Adaptation process	Av1000	NumCC	GtoH	FracH	FracL	Rstd
2DK	1FGA	2.025 (.069)	30	127	.213	.172	0.313 (.045)
	BMGAS	2.554 (.009)	30	26	.998	.000	0.075 (.006)
	ImitS	2.213 (.352)	15	27	.868	.000	0.016 (.003)
FRNE	1FGA	2.035 (.089)	30	122	.226	.161	0.292 (.049)
	BMGAS	2.572 (.007)	30	26	.996	.000	0.077 (.007)
	ImitS	2.176 (.507)	14	26	.901	.000	0.017 (.014)
FRN	1FGA	1.884 (.120)	30	162	.182	.311	0.379 (.044)
	BMGAS	2.476 (.026)	30	40	.939	.003	0.124 (.062)
	ImitS	1.362 (.434)	3	12	1.00	.000	0.009 (.004)
Tag	1FGA	2.198 (.057)	30	80	.380	.064	0.248 (.058)
	BGMAS	1.613 (.277)	30	331	.117	.499	0.469 (.093)
	ImitS	1.573 (.187)	0	-	-	-	0.012 (.004)
2DS	1FGA	1.484 (.086)	30	695	.040	.764	0.273 (.080)
	BMGAS	1.089 (.003)	1	978	.001	.977	0.024 (.009)
	ImitS	1.096 (.013)	0	-	-	-	0.006 (.000)
RWR	1FGA	1.502 (.109)	30	612	.055	.729	0.332 (.064)
	BMGAS	1.098 (.036)	9	1384	.021	.883	0.063 (.097)
	ImitS	1.104 (.032)	0	-	-	-	0.006 (.001)

The rules of the spatial game are as follows. Each player occupies a lattice point on a two-dimensional lattice space[9] and competes with all adjacent players (Moore neighborhood, 8 neighbors placed vertically, horizontally, and diagonally; see Fig. 6.3). The score of this matchup is summed and the payoff is calculated. Depending on the gain, the players at each lattice point change their strategies to the highest scoring strategy of the opponent with whom they fought. The following payoff table is adopted here:

$$\begin{array}{cc} & \begin{array}{cc} C & D \end{array} \\ \begin{array}{c} C \\ D \end{array} & \begin{pmatrix} 1 & 0 \\ b & 0 \end{pmatrix}, \end{array} \tag{6.9}$$

where $3/2 < b < 5/3$.

In addition, the cell (squares) of each player is colored in the following way:

■ Blue: C in the previous generation, C in the current generation

■ Red: D in the previous generation, D in the current generation

■ Green: D in the previous generation, C in the current generation

■ Yellow: C in the previous generation, D in the current generation

---

[9]The top and bottom, and the left and right of the square are connected, i.e., it has a toroidal structure.

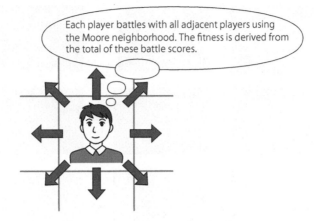

Figure 6.3: IPD as the spatial game.

$$C \quad D$$
$$\begin{matrix} C \\ D \end{matrix} \begin{pmatrix} 1 & 0 \\ b & 0 \end{pmatrix}$$
$$3/2 < b < 5/3$$

Figure 6.4: Walker.

Therefore, blue and red indicate a cell of a player who has not changed, and green and yellow indicate a cell that changed. The more green and yellow cells there are, the more changes occurred.

For example, let us take a look at the arrangement in Fig. 6.4. Here, the red cell indicates All-D and the blue cell indicates All-C. In other words, it is a cluster of 10 collaborators (All-C) surrounded by traitors. Numbers and symbols in cells are the total payoff obtained from the matches against the Moore neighborhood. For example, the value 5 for the blue at the base of the arrow is given as follows:

$$\text{5 matches against All-C} + \text{3 matches against All-D} = 5 \times 1 + 3 \times 0 = 5$$

Let us change the color of the cells one after another by executing IPD. Then this cluster (walker) moves in the direction of the arrow in the figure. In other words, in each generation legs move alternately: Right, left, right, left,... and it looks as if it were walking with two legs.

Figure 6.5 shows the kaleidoscope obtained by evolution, Fig. 6.6 shows the evolution leading to the Big Bang. A collaborative Big Bang is born by the collision of the walkers (Fig. 6.4). A kaleidoscope is produced by a single traitor (All-D)

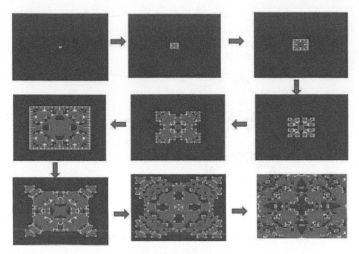

Figure 6.5: Kaleidoscope.

Color version at the end of the book

that invades a fixed size square collaborator (All-C). The expansion of a constantly changing symmetric pattern appears for a surprisingly long time. Since the total number of possible placements is finite, the kaleidoscope reaches a fixed pattern or cycle after a long time.

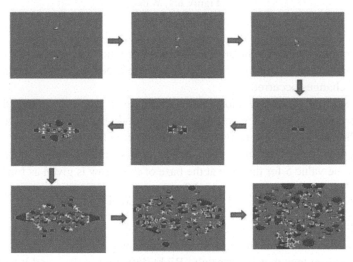

Figure 6.6: Big Bang.

There are also studies that extend this space model by incorporating asynchronous elements [52]. The game was changed to a procedure in which first a selection is made from matchups of several neighborhood groups, after which another selection is made from a matchup with a different group. As a result, it has been observed that in the end all planes were occupied by traitors.

IPD has been extended in various ways, and it is actively studied in fields such as artificial intelligence, artificial life, evolutionary economics, etc. Some of them will be explained below.

6.1.7   Can the Dilemma be Eliminated with Quantum Games?

Research has been made to solve the dilemma based on quantum games. This is an attempt to explain the seemingly irrational human behavior by considering a conditional strategy. Let us make Alice and Bob the players of the Prisoner's Dilemma game. The payoff table in Table 6.8 was used. Alice predicts Bob's behavior and decides her action. When she is sure that Bob will betray her, she will most likely cheat as well. On the other hand, even when Bob chooses to cooperate, Alice will betray him with a certain probability. According to psychological experiments, she will betray with 80% probability (cooperates with 20% probability). When Bob's next action is unknown (probabilities of his betraying or cooperating are equal), the chance to betray in classical theory is $90\%(=(80+100)/2)$ (called "natural law"). However, according to psychological experiments, the probability of betrayal of subjects is about 40%. This point was difficult to explain with the conventional logical reasoning.

Table 6.8: IPD payoff matrix and its code.

	Opponent (P_2) does not confess $P_2 = C$	Opponent (P_2) confesses $P_2 = D$
Oneself (P_1) does not confess $P_1 = C$	payoff : 3 Code : R	payoff : 0 Code : S
Oneself (P_1) confess $P_1 = D$	payoff : 5 Code : T	payoff : 1 Code : P

Here, a strategy called quantum strategy Q will be introduced. This is a quantum superposition of cooperation and betrayal at the degree. In other words, before Alice and Bob choose a strategy, they make each state entangled (quantum entanglement) and inform them about that, after which they select a strategy. At this time, it is shown that quantum strategy Q is a Nash equilibrium point and pareto efficient. This eliminates the Prisoner's Dilemma. Although the natural law is not satisfied, it can be explained as a probability theorem of quantum physics. In other

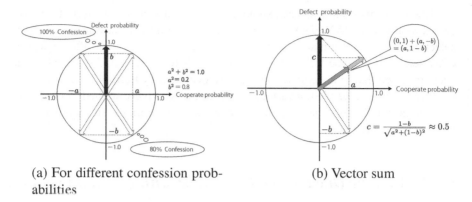

(a) For different confession probabilities

(b) Vector sum

Figure 6.7: State vectors for prisoner's dilemma by quantum computing.

words, acts of betrayal with 80% and 100% probability are independent rational judgments, but the probability of cheating decreases as a result of interference between subjects. In Fig. 6.7, dimensions of the quantum state vector are assumed to be probabilities to remain silent and confess. In Fig. 6.7(a), a state vector with 100% probability to confess (black) and a state vector with 80% probability to confess are drawn. There are several state vectors with 80% chance to confess, which satisfy $a^2 + b^2 = 1.0, a^2 = 0.2$ and $b^2 = 0.8$. When the opponent's state is not known, the two vectors, black and white, are superimposed to create a new state vector. At this time, when the state vector pointing to the lower right is selected, it becomes as shown in (b). Normalizing the resulting vector and solving it for the probability to confess c, yields the following $c = \frac{1-b}{\sqrt{a^2+(1-b)^2}} \approx 0.5$, which provides results close to the psychological experiment presented above.

Furthermore, considering the quantum games, it has been shown that cooperative behavior evolves more easily even on network structures, whereas, in the classical framework, the cooperation has been difficult to evolve [69].

The interpretation of quantum games described here does not mean that the human brain or intelligence is based on quantum computation. It is considered as a means of expressing the fluidity of thought that is called "Quantum Cognition." On the other hand, the physicist Roger Penrose, famous for the Penrose Tile, argued that our spirits can transcend ideas based on reason, so it cannot be reproduced with machines [88], and that the quantum mechanical properties are involved in the problems of behavior and consciousness of the brain on the macro scale, thus, proposing the "quantum brain theory" [89].

6.1.8 The Ultimatum Game: Are Humans Selfish or Cooperative?

The basis of the Prisoner's Dilemma is that humans are inherently selfish, that is, usually they are supposed to betray. This theory is based on the idea of "selfish genes[10]" advocated by Dawkins[11]. According to this, altruistic behavior and group selection will be wrong. Rather, the selfish behavior of the gene causes a non-selfish motive (a gentle, deep altruistic mind [91]) in the human brain.

However, is mankind not really considerate? Regarding this point, recent research results of the ultimatum games have raised a big question. The ultimatum game is defined as follows (Fig. 6.8):

> **Ultimatum game**
> There are two players A and B. They are not being informed of one another yet. They are given a chance to share a fixed amount of money and have only one chance. A is handed over 20$ and presents an amount between 0 and 20$ to B. B decides whether to accept or decline the amount presented by A. When B accepts, they share money as proposed by A and if B refuses, they both get nothing. What is the optimal strategy for A at this time?

The economically optimal strategy for this is clear. One dollar (i.e., the minimum unit for the proposal) should be better than nothing. Therefore, B should accept the proposal even if it is only one dollar. From this, A should propose to give only 1 dollar and keep the rest for himself. However, ordinary people do not think in this way. Usually, B declines when presented with a lower amount. B expresses his anger even at the price of missing money when he is clearly being taken advantage of. On the other hand, A usually does not offer such an unfair proposal. A proposes half of the money received to B. To avoid the proposal to be turned down, people often behave in a generous way. For example, the answers obtained by the author[12] are shown in Fig. 6.9. In this Japanese version of the question, the amount of money presented to A is 10,000JPY (Japanese Yen). 1US$=109.4270JPY (as of Jan. 31, 2019). The minimum unit for the proposal is 1JPY.

---

[10]The idea that genes are skillfully acting to succeed in breeding and leave as many descendants as possible. We are considered to be machines for gene survival (i.e., vehicles), living to make genes in our bodies survive. Many biological phenomena are explained by this way of thinking.

[11] Richard Dawkins (1941–): English evolutionary biologist. He has written many general books and general introductions to biology and, as a result, espoused thinking on the "Selfish gene," "Meme" (cultural information replicators), and "expanded phenotypes" (host operation due to the parasites, dams made by beavers, and mounds of white ants can be seen as phenotypes). His revolutionary ideas and provocative comments about evolution are still causing many discussions. He is a famous atheist.

[12]Data summarized in the same class as Table 4.2. However, this data is the total of students attending three lectures in 2014-2015.

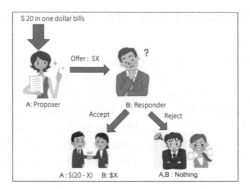

Figure 6.8: Ultimatum game.

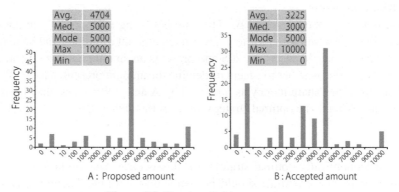

Figure 6.9: Results of ultimatum game.

It is surprising that science students who should be good at logical thinking are losing money due to their morals. The median value of A's offer is 5,000JPY, while the mode of B's accepted amount of money is 5,000JPY. However, the existence of rationality is somewhat seen, thanks to the fairly frequent appearance of 1JPY as B's accepted amount of money.

As a stricter game, there is the following game called a Dictator game (Fig. 6.10).

Dictator game

There are two players A and B. They are given a chance to share a certain amount of money and have only one chance. A is handed 20$ and gives the preferred amount of money between 0 and 20$ to B. Only A (=dictator) decides the result. What is the optimal strategy for A at this time?

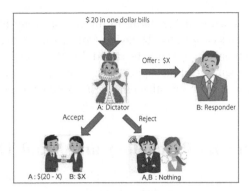

Figure 6.10: Dictator Game.

As a result of experiments conducted in many areas (Chicago, West Mongolia, Tanzania), even in this kind of ultimatum game, A (the dictator) shared a lot of money. It is said that on average 4$, i.e., 20% of the whole amount, was given to the opponents [68].

These results shook the foundations of standard economics. The foundation of modern economics is based on Homo Economicus (Economic man: Ultra-rational and egocentric person). The above result means that "economic man" as described in the traditional economic textbook is non-existent. In other words, compassion and cooperative behavior may originally exist in human beings.

Furthermore, when a research group inhibited the function of the right dorsal prefrontal cortex of the brain using electric stimulation during the ultimatum game, subjects were driven by greed and accepted a small amount of money, even though they kept thinking that it was unfair [32]. It seems that this area of the brain usually suppresses self-interest (accepting any money) and prevents selfish desires from influencing the decision-making process. This area is extremely important for taking fair action. In fact, there is a report that those who have damaged this area when they were young, cannot stand in the position of others, even after becoming adults, and they take up self-centered attitudes whenever confronted with ethical problems. In other words, this area suppresses selfish reactions and seems to play a major role in acquiring social knowledge.

In addition, when oxytocin[13] is sprayed on the subject's nose, that person seems to trust his/her partner even more [43]. As a result, when an experiment of Prisoner's Dilemma game was conducted, in which teams were formed to compete, the subjects sprayed with oxytocin tried to contribute to their team without making a selfish decision. They also acted more actively to interfere with the members of the other teams and to protect their teams. However, oxytocin does not bring about

---

[13] A hormone synthesized in the paraventricular nucleus of the hypothalamus and neurosecretory cells of the anterior nucleus. It is secreted from the posterior pituitary. The hormone is also called "love potion" and "affection hormone" because it relieves stress and brings a happy mood.

altruism. It raises a sense of belonging to the social group to which the subject belongs, and promotes cooperative behavior. Conversely, aggressiveness to people outside of the social group appears to be raised. In other words, the neurobiological mechanism of oxytocin secretion supports the view that it evolved to promote and maintain harmonious and cooperative relationships in a closely-related group.

6.2   Evolutionary Psychology and Mind Theory

Evolutionary psychology is a field aiming at a biological explanation of human behavior. It was founded by anthropologist John Tooby, psychologist Leda Cosmides, and many other biologically-oriented anthropologists. They objected to the idea of viewing the brain as a general-purpose learning machine. They claimed that each module of the brain existed to be functional and that its design was the result of natural selection. Therefore, the human mind is thought to have evolved to adapt to past environments [99]D

Tooby uses the analogy of the army knife. The army knife consists of screwdrivers and blades specialized for different functions. Similarly, there are visual modules, language modules, sympathy modules, etc., in the brain[14]. The brain is equipped with these innate means of data processing [99].

Evolutionary psychology focuses on the human mind so that it is distant from social biology that explains actions by focusing on genes[15] [109]. In particular, Tooby states that the theory of natural selection itself is an elegant inference device. Through this theory, a chain of deductive thoughts can be explained [13].

Evolutionary psychology has succeeded in explaining human cognition and mental mechanisms in terms of evolution. The representative research results were presented by an evolutionary linguist, Steven Pinker [90]. He claimed that language was rooted in human nature and that communication evolved through ordinary natural selection. That is, the language is a highly specialized organ. Examples of people acquiring sign language, etc. are raised as evidence, that the language is an instinct (language instinct) [92]. In other words, the language is not simply a product of culture, but an innate product unique to humans.

Since the history of evolutionary psychology is still shallow, while a variety of studies have been published, some of which have been overwritten by new experimental results. Also, there are many who object to the evolutionary biological view of humanity. For example, Steven Jay Gould has criticized that evolutionary

---

[14]On the other hand, the neuropsychologist Elkhonon Goldberg advocates the idea of a brain with gradient-type working principle. Modularization is true for the thalamus, but the cerebral cortex is a gradient-type and it is said to complement the modular approach [35].

[15]A research field to study the social behavior of animals based on the results of biology. Edward Wilson, a proponent of social biology [124], included a human in the last chapter of his book "Man: From Sociobiology to Sociology." Regardless of his intentions, he was criticized as an eugenicist and debate on social biology came about. The controversy gradually escalated emotionally and water was poured on Wilson during one of his lectures.

psychology explains everything only with adaptation, many of which are just afterthoughts of the "what and why story" type [90]. In this section, let us explain recent topics on Prisoner's Dilemma based on evolutionary psychology.

Economists Andrew Schotter and Barry Sopher [73] verified the cultural succession process in a game similar to the Prisoner's Dilemma called "Battle of the Sexes". In this game, the couple fights on whether to go to opera or soccer. They do not want to go separately, but the destinations are different. There are two Nash equilibrium points in this game (see Table 6.9). In other words, this is an unfair cooperative game where there are no optimal and equally satisfying solutions for the two. Such games are often found in everyday life. Andrew Schotter and Barry Sopher have experimentally verified whether a tradition arises in this game and what kind of cultural succession process it produces. As a result, the process of cultural evolution similar to a punctuated equilibrium process[16] was observed.

Table 6.9: Battle of the sexes.

		Husband	
		Opera	Football
Wife	Opera	2, 1	0, 0
	Football	0, 0	1, 2

◯ : Equilibrium point

Interestingly, there are cases where the prisoner's dilemma disappears [61]. The upper part of Fig. 6.11 is a standard prisoner's dilemma. The lower part shows that the payoff matrix changes if the two prisoners are siblings sharing half of the genes. Half of the payoff of a sibling prisoner B is added to prisoner A's payoff. Likewise, half of the payoff of the prisoner A is added to the prisoner B. At this time, the best policy for each is to cooperate regardless of what the other party chooses. If the situation of close relatives is taken into account in the process of evolution, there is a good chance that the cooperative behavior will emerge.

It is known that human beings have cognitive behaviors that cannot be explained with rational thinking. Self-deception and cognitive dissonance[17] are representative examples. It may be possible to explain this mysterious cognitive behavior from the point of view of evolution. For example, according to the zoologist Robert Trivers

---

[16]A hypothesis advocated by Stephen Jay Gould and Niles Eldredge. The idea that evolution occurs intensively in a short period of time when species diverge. Neutral gene recombination frequently takes place during long stagnation. Some kind of trigger, such as a change in the environment, causes a rapid evolution. There is a maxim that stagnation is data.

[17]An unreasonable behavior to convince oneself, in order to eliminate uncomfortable feelings with inconsistent perception. For example, the reason why a person cannot quit smoking even though the person knows that it is bad for health, is the belief that continuous smoking helps with mental stability or is a diet.

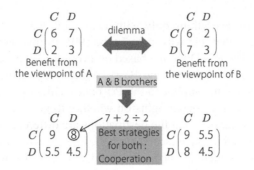

Figure 6.11: Disappearance of the prisoner's dilemma.

and Steven Pinker, adaptation to cooperation is said to have generated moral emotions [93]. In addition to that, exaggerations on one's kindness and skills can be seen through, thus, further improving skills [90]. In other words, a psychological arms race (co-evolution) occurs, such as one between liars and lie detectors. There is a proverb saying "Liars have to be quick to learn (Liar should have good memories.)". Triggers for the lies to be found out are external clues such as inconsistencies, contradictions, hesitation, twitching blushes, sweating, etc. In order to avoid such external cues, Trivers's hypothesis is that some degree of self-deception evolved through natural selection. If you lie to yourself, you will be more likely to be trusted by others.

Although it is challenging, there is an evolutionary psychology explanation for religion. For example, Richard Dawkins, known as an atheist, states that religion is a virus [25]. He insists that sometimes religion is harmful and its alternative, atheism should be socially accepted.

Also, religion is not a single subject. What is called a religion in modern western societies is a culture with another set of laws and customs, which survived alongside the laws and customs of the nation by accident. Religion has created art, philosophy, and law, and it has made a profit for those who advertised it. We conclude this section with Pinker's quotation about religious emergence, which solves the problem of the afterlife (see page 161).

6.3   How does Slime Solve a Maze Problem? Slime Intelligence

Figure 6.12 shows a simulator of bacterial colonies based on DLA. In this system, the eating speed of bacteria, movement speed, the amount of food necessary for the division, the diffusion rate of nutrients, etc. are variable. The distribution of bacteria and distribution of nutrients are shown in another window, and the number of bacteria and the total amount of nutrients are plotted on graphs. When an exper-

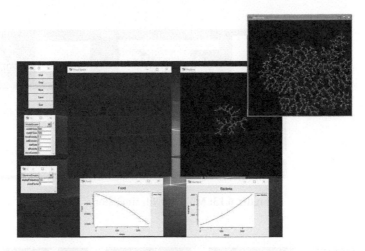

Figure 6.12: Bacteria colony simulator.

iment was conducted, the bacteria were observed to form circular colonies when sufficient nourishment was available. However, when the density of nutrients was low and the nutrients needed for the division were small in comparison to the eating speed, the colony formed a pattern as shown on the right. This is considered to be a strategy to search for nutrients in a wide area using the least amount of bacteria in an environment with scarce resources.

As a more practical simulation, let us consider the growth of red bread mold (Neurospora crassa). How mold grows is described in detail in the literature [101]. Sufficient nutrition, oxygen, humidity, and appropriate temperature are necessary for mold to grow. When these conditions are met, the mold grows and multiplies by producing spores that are released into the air. The released spores usually fall to places unsuitable for growth and die. However, spores that luckily landed in places satisfying the conditions survive. When the spores survive, they grow, produce sporangia, and new spores are released into the air. Here, the simulator modeling mold growth followed the rules below:

■ There must be enough oxygen for mold to grow.

■ The humidity must be between 70 and 90 percent.

■ Temperature must be between 10 and 35 degrees.

■ The mold floating in the air hits the surface at a certain rate.

■ After a specific step, mold begins to emit spores.

■ Whether the mold attached to the surface grows, depends on temperature, humidity, and characteristics of the surface.

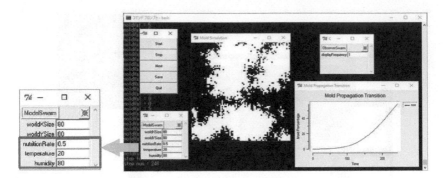

Figure 6.13: Mold growth simulator.

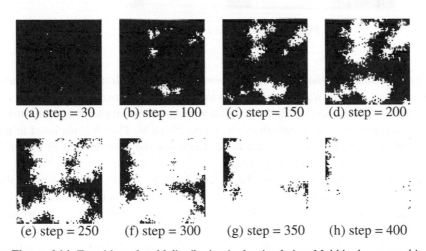

(a) step = 30   (b) step = 100   (c) step = 150   (d) step = 200

(e) step = 250   (f) step = 300   (g) step = 350   (h) step = 400

Figure 6.14: Transition of mold distribution in the simulation. Mold is shown as white dots.

Although, strictly speaking, these are not DLA, they can be thought to be an extension. Figure 6.13 shows an overview of the simulator. Nutrition rate, humidity, and temperature can be set in this simulator. The growth of the mold is drawn on two dimensions. In addition, the occupancy of the mold is plotted and displayed every hour.

Figure 6.14 shows the growth of the mold every 100 steps. In the figure, the mold is rendered in white. Almost every place was occupied by mold in about 500 steps. The parameters at this time are as shown in Table 6.10.

During the simulation, the mold seems to create groups (clusters) and grow. When spores spread out from the current mature mold, they fall down near the mold that produced them and grow more easily. Figure 6.15 shows the transition of the mold occupancy rate in time in the simulation. This coincides well with actual

Table 6.10: Experimental parameters (growth of Neurospora crassa).

Parameter	Value
world X size	80
world Y size	80
nutrition rate	0.5
temperature	20 [°C]
humidity	80 [%]
oxygen	true
mold emerge rate	0.001
mold growth step	5

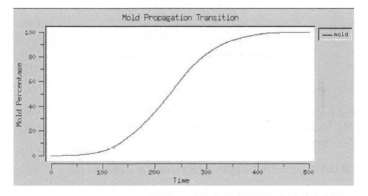

Figure 6.15: Percentage of mold in the simulation.

results of mold observation[18]. In favorable conditions, mold grows in three stages. It multiplies very slowly at the beginning. After occupying 10% to 20% of the space, the growth speed dramatically increases. However, the growth rate declines when the mold occupies 80% to 90% of the space. This phenomenon is thought to occur from the balance of the number of fungi scattering new spores to the free areas and the space occupied by new spores.

Slime mold (Myxomycete) are small creatures living in fallen trees and soil and have two forms, one that moves like amoeba and one that looks like a mushroom. In other words, a trophozoite called plasmodium is a unique living organism, possessing both the animal nature that feeds on microorganisms while it is moving as well as the vegetable nature of breeding through spores. The life cycle of the mold is shown in Fig. 6.16.

---

[18]For example, refer to the experimental results (condition A) shown in `https://explorable.com/mold-bread-experiment`.

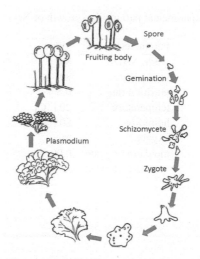

Figure 6.16: The life cycle of slime mold.

A study has been conducted to solve the shortest path of a 2-dimensional maze using slime mold [81]. This search method using slime mold consists of the following two steps:

Step1 Fill the maze with the plasmodium of the slime mold.

Step2 Place sources of nutrition at the entrance and exit.

In **Step**2, the shortest path is obtained by optimizing the shape in which the slime mold incorporates nutrition. This is made possible by the interactions of contractive movements by rhythmic pulse present in slime mold and morphogenesis of the tubes transporting nutrients. It is surprising that primitive organisms, such as slime mold, exhibit intelligent behavior, that is, the simple behavioral principles of the slime mold cells solve advanced problems. This related study has won the (cognitive science field) Ig Nobel Prize in 2008 (see Fig. 6.17).

Nakagaki et al. modeled the nutrient flow made up of trophozoite of the slime mold constituting the tube-shaped flow path from the fluid dynamics (hydrodynamics) point of view and realized a method of performing a route search as a physical simulation.

The basic behavior of the Physarum polycephalum (many-headed slime) consists of morphological and ecological behaviors.

The following are known as morphological properties:

1. When scattered trophozoite come in contact with each other, they easily fuse and form a single trophozoite.

2. Macroscopically, they consist of finely-branched, tube-shaped (tubular) flow path networks.

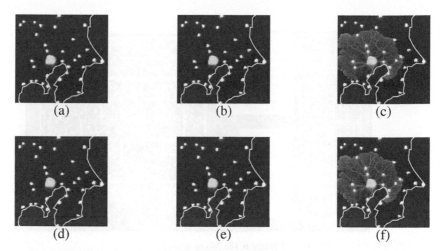

Figure 6.17: Network formation by slime intelligence. Slime molds successfully formed the pattern of a railway system quite similar to the Japanese railroad networks centering Tokyo [116].

3. They try to maintain individuality as much as possible (prefer to bind together rather than dividing).

4. They try to surround a source of nutrition with their big bodies.

Also, there are the following ecological properties related to the route search:

1. Tubes in (parallel to) the direction of contraction movements of the rhythm cells are strengthened, and perpendicular/orthogonal tubes fade.

2. The faster the rhythm of the contraction movement is, the more the cell is supplied with nourishment.

In the following simulation, the above-mentioned slime maze route search will be reproduced with a single slime mold cell as an individual. Individuals connected to each other are distributed throughout the maze beforehand. The lifetime of the individuals is assumed from the transportation and ingestion of the nutrients. In other words, attention is paid only to the properties of the cells to intentionally approach and bind to the food or cells that absorbed the nutrition. At this time, each cell has a certain field of view (5 cells) in the horizontal or vertical direction, and if a cell neighboring the food or food itself is detected in the field, they approach it. The whole field, the distribution of cells, etc. can be adjusted with the control panel (see Fig. 6.18).

Initial values were set to 5,000 slime mold cells distributed over a square maze with a width of 124 dots, and the result is shown in Fig. 6.19. The yellow dots in the figure are mold, and the two red dots are the starting point and the goal of the

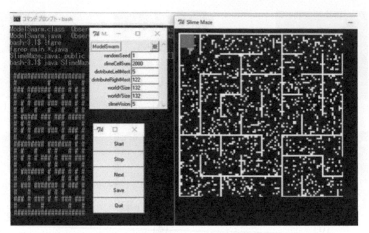

Figure 6.18: Maze search simulator.

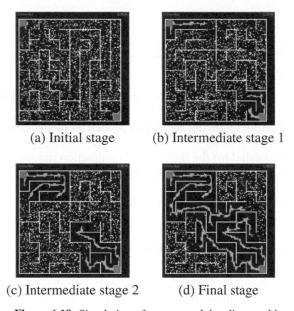

(a) Initial stage   (b) Intermediate stage 1

(c) Intermediate stage 2   (d) Final stage

Figure 6.19: Simulation of maze search by slime mold.

maze. In the initial state (a), yellow slime mold cells are widely distributed. Out of those, the mold cells close to the food in red squares try to come into contact with the food. Also, the orange cells that came into contact with the food, become targets of other yellow cells. As a result, an intermediate state (b) can be observed, where cells are in progress on the path. Additionally, as the stage proceeds, the pathway branching occurs and cells merge, thus, the maze is explored (see the intermediate stage (c)). In the final situation (d), the cells grow until the connection between the starting point and the goal is formed, providing a solution of the maze.

The simulations described above are based on DLA, and the only rule followed by each cell is to come into contact with an element containing nutrition. However, even the simplest rule, such as the one used here, enables exploration of the path out of a simple maze. But the temporal and spatial costs of the route search are too high (bad) compared to the other maze-solving algorithms. For more practical simulation of a slime mold, it is necessary to add additional information related to the physiological properties of slime mold, such as cell generation and destruction, the principle of tube formation, and the transport of nutrients.

6.4 Swarm Robots

6.4.1 Evolutionary Robotics and Swarm

Evolutionary robotics [84] is a method for producing autonomous robots via evolutionary processes. In other words, it is an attempt to train appropriate robot behavior by representing the robot control system as a genotype and then applying EA search techniques.

Evolutionary robotics was proposed in 1993 by Cliff, Harvey, and Husbands [19], but evolutionary simulations of artificial objects with sensor motor systems were being conducted in the latter half of the 1980s. At that time, the control of production robots required special skills and advanced programming abilities. In a clear departure from traditional robotic design, a new generation of robots appeared in the 1990s that shared features of simple biological systems. These features are robustness, simplicity, compactness, and modularity. Those robots were designed to be easily programmed and controlled, even by people with backgrounds quite different from that of a conventional robot technician.

For example, Floreano et al. [84] have verified the Red Queen effect[19] through an evolutionary experiment with two Khepera robots (predator and prey). Each robot has six infra-red sensors at the front, and another two at the back. The predator is also equipped with visual sensors. The maximum speed of the prey is twice that of the predator. If the prey can run away cleverly, it will never be caught by the predator. Against this, the predator has more sensor information available to track the prey. The behavior of the predator and the prey were represented as genotypes via evolutionary ANN[20] direct coding and coevolved. One of the indexes used to characterize fitness was how well the prey survived. In other words, the longer the prey can evade capture by the predator, the greater the prey's fitness. Conversely, the faster the predator can catch the prey, the greater its fitness. The result of co-

---

[19]Red Queen effect is an evolutionary type of arms race. The name is derived from Alice's Adventures in Wonderland. It is a model of co-evolution.

[20]EANN, see [54] for details.

evolution was the successive appearance of a series of fascinating tracking and evasion maneuvers, just as predicted by the Red Queen hypothesis. Examples of these behaviors include the "spider strategy" (backing away slowly until a wall becomes visible and then catching the prey as it approaches) and circling one another like a dance.

In recent years, evolutionary robotics has been the focus of attention from researchers in a variety of fields, including artificial intelligence, robotics, biology, and social behavioral science. In addition, evolutionary robotics shares many features with other AI approaches. These include:

■ behavior-based robots,

■ robot learning,

■ multi-agent learning and cooperative learning,

■ artificial life,

■ evolvable hardware, and

■ swarm robotics.

Swarm robotics is a field of multi–robotics in which a large number of robots are coordinated in a distributed and decentralised way according to swarm intelligence. The main source of inspiration for swarm robotics comes from observing the behavior of social animals and, in particular, social insects such as bees, ants and termites (see Ch. 3 for details). Studies have revealed that no centralized coordination mechanisms exist behind the synchronized operations in social insects, and yet their system's performance is often robust, scalable and flexible. These properties are desirable for multi–robot systems and can be regarded as motivations for the swarm robotic approach [12].

The robot systems can be separated into two categories: Centralized and decentralized systems.

In centralized systems, a single camera observes the environment and the robots as a whole, and a central computer processes the camera information to remotely control each robot in order to accomplish a task. Another important characteristic is that they possess a leader, which guides the rest of the robots. This leads to several limitations, e.g., the leader's death results in the loss of the team's guidance, etc., which means to be not robust in real-world applications.

The decentralized systems operate without a leader, leaving the decision making and the completion of certain tasks to each robot as individual. Given this, the robots should have enough sensors (such as cameras, distance sensors, etc.) in order to obtain sufficient information from its surroundings and produce a similar behavior as found in nature [117]. The advantages of these systems are robustness, scalability, flexibility and collective behavior.

In this section, we will explain how swarm robotics work for a transportation problem in the decentralized way.

6.4.2 Transportation Task for Swarm

The transportation of significantly large objects by swarm robots has been studied since the industry started to develop. Engineers [56, 118] and researchers [3, 118, 117, 120, 127] have done plenty of work on this topic. For the task described in [127], two large robots were used. However, these solutions tend to become expensive. Nowadays, we can obtain computationally powerful robots at an affordable price with good hardware quality. On the other hand, for the sake of transportation, conventional robot hardware does not offer enough power and torque. Nevertheless, researchers have developed the idea of using swarm intelligence on low-cost robots in order to overcome these limitations.

The object transportation task using autonomous systems has been approached and applied in many works. A collective transport strategy, inspired by the food retrieval procedure of ant colonies, has been implemented on a swarm of robots that are smaller than the object. The three most common types of strategies are pulling, pushing, and caging [15].

In [72], the transportation of multiple objects dispersed in the environment is studied. A non-prehensile manipulation method with a single robot was used, putting emphasis on the path planning and task allocation.

The transportation of an object was carried out by a multi-robot optimization [103]. In this research, two robots jointly carried a stick from an initial position to a final destination, without colliding against the obstacles. The sensory data of the robots are the input variables of the optimization problem, while the output variables are the necessary amount of rotation and translation of the stick.

Alkilabi et al. [3] presented a set of simulation experiments, in which autonomous robots are required to coordinate their actions in order to transport a cuboid object too heavy for a single robot to move.

6.4.3 Occlusion-based Pushing (OBP)

Occlusion-based pushing [15] consists of pushing an object as long as the robot cannot see the goal (or the source of light or color). The goal is the immediate place where the robots have to push the object (a red circle in the environment, see Fig. 6.23). Because the robots have no specific ways to grasp the object, pushing is the best way to move the object.

A positive aspect of this transportation strategy is that, rather than being treated as a problem to solve, occlusion is utilized to organize a swarm of robots to push a large object toward a goal. The algorithm deals with pushing the object across the portion of its surface, where it occludes the direct line of sight to the goal. This may result in the transportation of the object along a path that is not necessarily optimal, but the object is always moved to the goal. Figure 6.20 illustrates how these robots work together by OBP method.

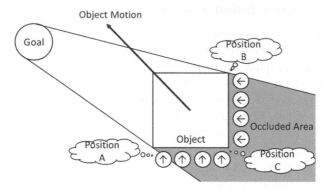

Figure 6.20: Illustration of how a swarm of robots can push a large object in a 2-D environment [15].

The simplicity of this method makes it particularly suitable for the implementation on mobile robots that have limited capabilities. In the long term, this simple multi–robot strategy will be realized at very small scales and for such important tasks as rescue scenarios.

Figure 6.21 presents a State diagram representation of the OBP algorithm proposed by Chen [15]. This diagram produces the following robot behavior:

■ Once the object is found, the robot moves toward it ("Approach Object").

■ When the robot has reached the object, it enters the state of "Check for Goal" to check whether the goal is visible from the robot's position.

■ If the goal is not visible, the robot will push the object simply by moving against it ("Push Object").

■ If the goal is visible, the robot will attempt to find another position around the object ("Move Around Object").

In order to estimate the effectiveness of the OBP algorithm, we have implemented a decentralized swarm system. For the sake of simulation, we use V-REP[21] as our rendering tool. The chosen mobile robot is the Khepera III[22]. The device can be found in the simulator as a premade sample. The OBP method was implemented according to the diagram presented in Fig. 6.21.

---

[21]Coppelia Robotics. V–REP. http://www.coppeliarobotics.com/
[22]K-Team Corporation https://www.k-team.com/

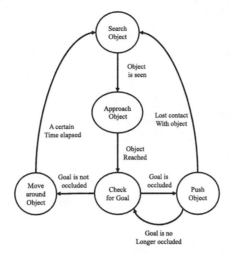

Figure 6.21: State diagram of a robot realizing the occlusion-based cooperative transportation [15].

Figure 6.22 shows the environment used for the first experiment (i.e., Environment 1). It is an area of $16m^2$ of working space with the object placed in the bottom-right corner. The goal is located in the top-left corner. We formed five groups of robots: Teams of 1, 2, 3, 6 and 12 robots, respectively. For each group, simulation was repeated five times in order to get the group's best time. The object has a mass of $1Kg$ for all trials.

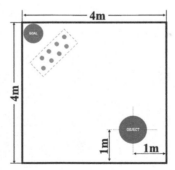

Figure 6.22: Environment 1: OPB experimental area. Goal is represented as a red circle and the object to transport as a blue circle.

In Fig. 6.23, we show the snapshots of robots' behaviors. The time is indicated as "minutes:seconds" with an interval of 30 second between the successive snapshots. Each horizontal series of snapshots corresponds to a different group of robots performing the OBP method. The number of robots is shown in the left side of the figure.

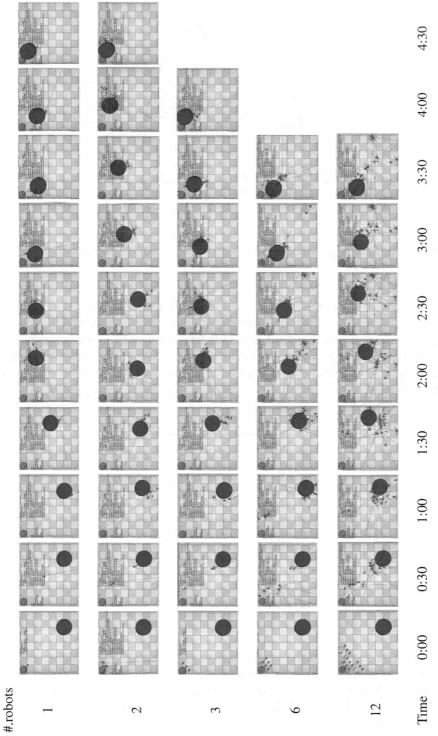

Figure 6.23: Swarm robots' behaviors by OBP algorithm (Environment 1).

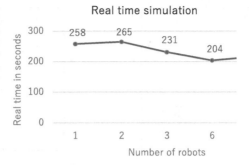

Figure 6.24: Best time for each group of robots.

For each group, all trials were successfully completed, but some runs took more time. In Fig. 6.24, we show the best time for each group. The best performance is given by a 6-robot team. This is because all the six robots are constantly working so as to push the object to the destination. In a 12-robot team, some robots tended to interrupt others.

6.4.4 Guide-based Approach to OBP

The traditional OBP only works in open spaces, where the robots can see the goal or the object after a 360-degree turn. However, if there are some obstacles between the object and the goal, additional self-organizing mechanisms are required.

Chen [15] presented two ways of pushing an object to the destination through an environment, where it is not possible to have a direct glance of the object and the goal within a 360-degree turn. The first idea is to manually place robots for the sake of forming a path with sub-goals; then, when some robot gets close to the object, this one will start to help the other robots push the object. The second approach is to cover one robot with the goal's color as a guide. Then, with a remote-control, the guide is moved through the environment until it gets close to the destination, while the rest of the robots are pushing the object. Because of the limitation of this method, i.e., the necessity of manual or remote control, we have proposed an algorithm that allows the swarm to push an object through a non-opened environment [66]. Note that the robots are assumed to have no information about the space surrounding them. Also, the robots do not have any ways to communicate between them.

Our algorithm allows the robots to produce a path, without their communication, so as to lead pusher robots where to bring the object back to the destination. This path will work like the pheromone trail of an ant community (see sec.3.1.1). A robot that takes up the role of creating a path is called a "guide."

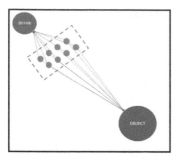

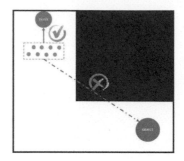

(a) Environment 1: Open space environment for OBP. Lines indicate that every robot can see the goal and the object.

(b) Environment 2: Non-open space. Lines indicate the limitation of the OBP method.

Figure 6.25: OBP experimental environments.

In our framework, each robot can take one of the following two modes:

- **Push mode**: A robot pushing the object by applying the OBP method (as shown in Fig. 6.21).

- **Guide mode**: A robot placing itself in a strategic position as a sub-goal for the purpose of leading other robots in Push mode.

When the traditional OBP condition is not fulfilled, our algorithm allows the robots to explore the area in order to decide whether to take a guide role or a pusher role (see Fig. 6.26). Guiding robots are necessary to let other robots retrieve the object to home successfully[23].

In the following part, Home is the place where the robots are deployed (represented as a red circle, see Fig. 6.25(b)). The difference between "Goal" and "Home" are as follows. The former is used in the original OBP method developed by Chen [15]. It is the immediate place where the robots have to push the object. The latter refers to the final destination of the object, which is the place where the robots are deployed. In our method, "Goal" will be a sub-goal represented by some guide robot.

The first step is to determine if the home and the object are visible within one 360-degree turn. If both targets are visible in the first check, the robot will perform Push mode (the left state in Fig. 6.26). If only the home is visible, the robot will play a guide role (the right state in Fig. 6.26).

---

[23]Note that this guiding strategy is robust because the guiding role emerges in a decentralized way. See the limitation discussion of the centralized robot system in page 198.

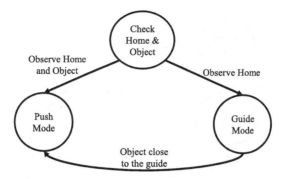

Figure 6.26: State diagram of the first stage of guided OBP method.

Guide mode has a state called "Explore." This state is performed by following a Braitenberg obstacle avoidance method [110], for the purpose of sorting the walls and robots in the environment until it can detect one of the targets (Home, guide or object). The other two states are "Light Color & Turn" and "Follow object". The former refers to the change of colors on the robot's external case from green to orange (color used to identify the guide robots). Then, the robot rotates on its own spot until the object becomes visible. The latter is performed after the object is visible from the robot. The robot takes small turns while observing the object, allowing it to keep the object perpendicular to its vision line.

The algorithm shown in Fig. 6.27 is executed in the following ways: The robot first explores the environment until it finds the object. Then, the robot approaches the object until it is touched. Afterwards, the robot checks if Home or a guide is visible from its position. If it is, the robot starts to perform Push mode towards the visible sub-goal. Otherwise, the robot explores again the environment until it finds Home or the closest guide. If the robot can see the guide or Home and also see another robot, it will go back to search for the object. If it can just see another guide or home, it will place itself at a specific distance to be watched by the other robots and form the path. When the robot is already placed in that position and is watching the approaching object, it will wait until the object gets close so as to change its state to Push mode.

When our method (Guided Pushing) is applied to a swarm of robots, it results in producing a path in the environment and, therefore, the rest of the robots learn to move the object, as if the robots use this guidance as a pheromone trail.

6.4.5 *Let us See How they Cooperate with each other*

In this section, we shall show how swarm robots work together by the proposed Guide-based OBP. The experimental set-ups are described below.

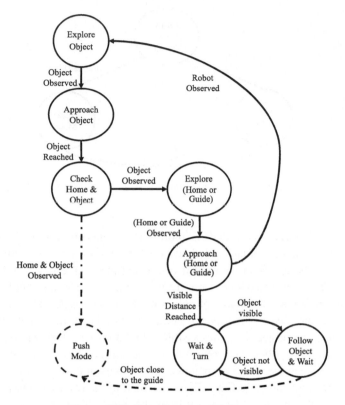

Figure 6.27: State diagram of Guide mode.

The robots can recognize four colors so as to differentiate the targets and other team members:

■ Red: Home representation.

■ Blue: Object representation.

■ Green: Normal robot representation.

■ Orange: Guide robot representation.

In order to show the effectiveness of our method, we have conducted experiments with V-REP simulator. The environmental design is based on the work of Chen [15] with slight modification. As shown in Fig. 6.28, we enlarged the environment to $64m^2$ with an obstacle of $25m^2$, resulting in a total working area of $39m^2$ with walls of $0.2m$ high. The object has a mass of $2Kg$.

In a larger and complicated environment, the traditional OBP method would be stuck at the "Search Object" (see Fig. 6.21). This is due to the impossibility of perceiving the object in a non-open environment. The robots can not locate and approach the object because of the obstacle of $25m^2$ in the area.

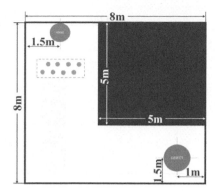

Figure 6.28: Environment 2: First experiment of Guided OBP method. Note that a red circle is Home.

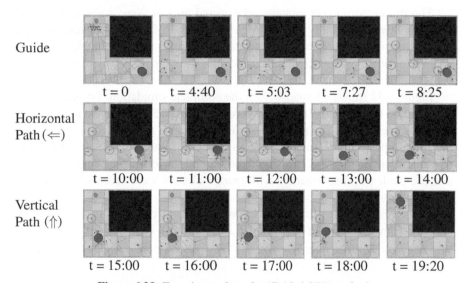

Guide				
t = 0	t = 4:40	t = 5:03	t = 7:27	t = 8:25

Horizontal Path (⇐)

t = 10:00   t = 11:00   t = 12:00   t = 13:00   t = 14:00

Vertical Path (⇑)

t = 15:00   t = 16:00   t = 17:00   t = 18:00   t = 19:20

Figure 6.29: Experimental results (Guided OBP method).

Color version at the end of the book

In Fig. 6.29, we show the snapshots of the simulation. The whole experiment took 19:20 (min:sec). Remember that the robots would not start the pushing process until they know where to push. We can observe that the time the last guide is placed (at the end of the guiding process) is 8:25 (min:sec), leaving the rest of the task to the pusher robots. When the object is close to one of the guides (see the snapshot of time 8:25 in Fig. 6.29), this one becomes a pusher that helps the rest in the process of retrieving the object. Overall swarm robots' motion in Guide mode is represented by arrow direction, i.e., ⇐ and ⇑.

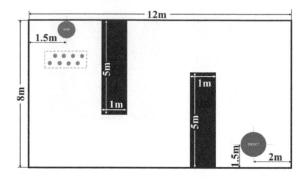

Figure 6.30: Environment 3: Second Experiment of Guided OBP method.

The third environment includes an external area of $96m^2$ and a total working space of $86m^2$ with walls of $0.6m$ high (see Fig. 6.30). These obstacles occupy $10m^2$ of the total area. Here, Guide mode algorithm is tested in this larger area, in order to observe how effectively the robots produce a path between the object and the home. The object has a mass of $2Kg$.

Figure 6.31 shows the snapshots of the complete simulation. The total duration was 2:40:00 (h:min:sec). The guiding process was 1:15:22 long and the rest of the time was spent on the pushing process. The snapshots are divided into the following three types: (1) the guide formation process in the first line, (2) the vertical movement in Guide mode, and (3) the horizontal movement in Guide mode. Note that the guide positions (marked with orange circles) were successfully identified by the pusher robots.

These results have demonstrated how the swarm of robots accomplished the task of retrieving an object in a complex and non-open environment.

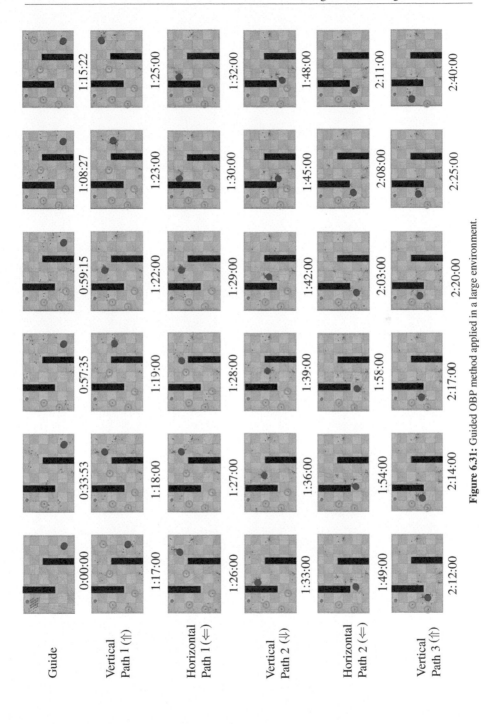

Figure 6.31: Guided OBP method applied in a large environment.

Chapter 7

Conclusion

Constantly making new combinations, and interesting combinations of ideas stick together, and eventually percolate up into full consciousness. That's not too different from a biological population in which individuals fall in love and combine to produce new individuals.

—Gregory Chaitin, META MATH! The Quest for Omega

7.1 Summary and Concluding Remarks

In this book, we have explained the concept of "emergence" for AI, focusing on swarm. The basic algorithm to achieve that is based on the theory of evolution.

As a final thought, we will explore the fact that evolution is still ongoing ("evolution at work").

The Galapagos Islands are famous for being the place where Charles Darwin came up with the theory of evolution in the middle of his voyage with the surveying British Navy ship Beagle. He had the idea after observing that the beaks of finch birds vary slightly in shape depending on the species, and that the shells of giant tortoises vary from island to island. However, this story is somewhat exaggerated. Darwin did not study these birds in depth in the field. Those who have read "On the origin of species" realize that the data from Galapagos was not fully utilized. Rather, his discussion was always restricted to artificial selection.

Nevertheless, this island continues to draw considerable attention as a core for research on evolution theory. For instance, Peter Grant from Princeton University conducted a thorough 20-year-long research on finches [39], clarifying that their beaks change dramatically due to selection pressure caused by food availability, which varies due to climate changes. Moreover, he found that evolution occurs at

(a) land iguana

(b) marine iguana

(c) hybrid iguana

Figure 7.1: Galapagos iguana.

an extremely fast rate (a matter of years), rather than the previously imagined period of tens of thousands of years.

Furthermore, the Galapagos Islands are home to two types of iguanas, namely, the land iguana and the marine iguana (see Fig. 7.1). The yellow land iguana originally lived on land and used to eat prickly pear cactus sprouts. However, co-evolution with cactuses resulted in lack of food. Eventually, arms races took place among cactuses, land iguanas, and giant tortoises. Cacti grew taller in order to protect their sprouts from being eaten. They also developed pricks in their stalks, to stop land iguanas from climbing on them. Thus, food shortage started to occur. It is conjectured that they evolved into sea iguanas that are able to swim and eat sea weeds. The lifestyle of sea iguanas differs greatly from that of land iguanas. Their bodies became black to better catch day light, and they became able to hold their breath and dive deep into the water. When they return to land, they expel salt from their noses and bask in the sun. An interesting fact that became a recent topic is the birth of hybrid iguanas, resulting from a female sea iguana and a male land iguana. They are pink-colored iguanas that exhibit dual characteristics. They also have nails that allow them to climb cactuses. However, at present they do not seem to be fertile.

As a way of thinking, evolution is not restricted to living beings with genes. The idea also holds in a variety of other systems. We can find several famous phrases and sayings about the matter. Matt Ridley [100] referred to the idea of evolution above as a way of thinking that can be widely applied as the "general theory of evolution." In contrast, the evolution of living bodies based on natural selection proposed by Darwin is considered a "special theory of evolution." As seen in this book so far, the swarming mechanism can explain several phenomena in society, currency, technology, language, culture, politics, and moral based on the general theory of evolution.

This book explains a variety of topics based on this idea. We hope that from now on, further developments based on evolution and emergence will occur in AI methodologies.

References

[1] Adamatzky, A., "Voronoi–like partition of lattice in cellular automata," *Mathematical and Computer Modelling*, vol.23, no.4, pp.51–66, 1996.

[2] Adamatzky, A., Costello, B., and Asai, T., *Reaction-Diffusion Computers*, Elsevier Science, 2005.

[3] Alkilabi, M.H.M., Lu, C., and Tuci, E., "Cooperative Object Transport Using Evolutionary Swarm Robotics Methods," in *Proc. of the European Conference on Artificial Life (ECAL)*, pp.464–471, 2015.

[4] Andre, D., Bennett III, F.H., and Koza, J.: "Evolution of Intricate Long-Distance Communication Signals in Cellular Automata using Genetic Programming," in *Artificial Life V: Proceedings of the Fifth International Workshop on the Synthesis and Simulation of Living Systems*, 1996.

[5] Arthur, W.B., *Increasing Returns and Path Dependence in the Economy*, University of Michigan Press, 1994.

[6] Auer, P., Cesa-Bianchi, N., and Fischer, P., "Finite-time analysis of the multi-armed bandit problem," *Machine Learning*, vol.47, no.2, pp.235–256, 2002.

[7] Axelrod, R.: "The Evolution of Cooperation," Basic Books, New York, NY, 1984.

[8] Axelrod, R.: "An evolutionary approach to norms," *American Political Science Review*, vol.80, no.4, pp.1095–1111, 1986.

[9] Bäck, T., Hoffmeister, F., and Schwefel, H.-P., "An Survey of Evolution Strategies," in *Proc. 4th International Conference on Genetic Algorithms (ICGA91)*, pp.2–9, Morgan Kaufmann, 1991.

[10] Bäck, T., "An overview of evolutionary algorithms for parameter optimization," *Evolutionary Computation*, vol.1, no.1, pp.1–23, 1993.

[11] Bellman, R., Adaptive Control Processes–A Guided Tour, Princeton University Press, 1961.

[12] Bonabeau, E., Dorigo, M., and Theraulaz, G., *Artificial Intelligence from Natural to Artificial Systems*, Oxford University Press, New York, 1999.

[13] Brockman, J., *This Explains Everything: 150 Deep, Beautiful, and Elegant Theories of How the World Works*, Harper Perennial, 2013.

[14] Civicioglu, P., and Besdok, E., "A conceptual comparison of the Cuckoo-search, particle swarm optimization, differential evolution and artificial bee colony algorithms," *Artificial Intelligence Review*, vol.39, no.4, pp.315–346, 2013.

[15] Chen, J., Gauci, M., Li, W., Kolling, A., and Groβ, R., "Occlusion-based cooperative transport with a Swarm of miniature mobile robots," *IEEE Transactions on Robotics*, vol.31, no.2, pp.307–321, 2015.

[16] Chu, S.-C., Tsai, P.-W., and Pan, J.-S., "Cat Swarm Optimization," in *Proc. Pacific Rim International Conference on Artificial Intelligence (PRICAI 2006)*, pp.854-858, 2006.

[17] Clarke, R.D., "An application of the Poisson distribution," vol.72, no.3, *Journal of the Institute of Actuaries*, 1946.

[18] Clerc, M., and Kennedy, J.: "The Particle Swarm: Explosion, Stability, and Convergence in a Multimentional Complex Space," *IEEE tr. Evolutionary Computation*, vol.6, no.1, pp.58–73, 2002.

[19] Cliff, D, Harvey, I., and Husbands, P., "Explorations in evolutionary robotics," *Adaptive Behavior*, vol.2, pp.72-110, 2000.

[20] Cohen, M.D., Riolo, R.L., and Axelrod, R.: "The emergence of social organization in the prisoner's dilemma: How context-preservation and other factors promote cooperation," Santa Fe Institute working paper 99-01-002, http://cscs.umich.edu/research/techReports.html, 1999.

[21] Couzin, I.D., Krausew, J., Jamesz, R., Ruxtony, G.D., and Franksz, N.R., "Collective memory and spatial sorting in animal groups," *Journal of Theoretical Biology*, vol.218, no.1, pp.1–11, 2002.

[22] David, P.A., "Clio and the economics of QWERTY," *The American Economic Review*, vol.75, no.2, Papers and Proceedings of the Ninety-Seventh Annual Meeting of the American Economic Association, pp.332–337,1985.

[23] Davies, N.B., Krebs, J.R., and West, S.A., "An Introduction to Behavioural Ecology," Wiley-Blackwell, 4th Edition, 2012.

[24] Dawkins, R.: *The selfish gene*, Oxford Univ. Press, 1991.

[25] Dawkins, R.: *The God Delusion*, Mariner Books, 2008.

[26] Deb, K.D., Pratap, A., Agarwal, S., and Meyarivan, T., "A fast and elitist multiobjective genetic algorithm : NSGA-II," *IEEE Transactions on Evolutionary Computation*, vol.6, no.2, pp.182–197, 2002.

[27] Devlin, K., *The Math Instinct: Why You're a Mathematical Genius (Along with Lobsters, Birds, Cats, and Dogs)*, Basic Books, 2009.

[28] Dewdney, A.K., "The hodgepodge machine makes waves," *Scientific American*, vol.259, no.2, 1988.

[29] Dewdney, A.K., "The cellular automata programs that create wireworld, rugworld and other diversions, Computer Recreations," *Scientific American*, 72, pp.146–149 Jan. 1990.

[30] Dorigo, M., and Gambardella, L.M.: "Ant colonies for the traveling salesman problem", Tech. Rep. IRIDIA/97-12, Universite Libre de Bruxelles, Belgium, 1997.

[31] Fung, K., *Numbers Rule Your World: The Hidden Influence of Probabilities and Statistics on Everything You Do*, McGraw-Hill Education, 2010.

[32] Gazzaniga, M.S., *Who's in Charge?: Free Will and the Science of the Brain*, Ecco, 2011.

[33] Geem, Z.W., Kim, J.H., and Loganathan, G.V., "A new heuristic optimization algorithm: Harmony search," *Simulation*, vol.76, no.2, pp.60–68, 2001. *Physical Review E*, vol.79, 2009.

[34] Ghosh, A., Dehuri, S., and Ghosh, S. (eds.), "Objective Evolutionary Algorithms for Knowledge Discovery from Databases," Springer, 2008.

[35] Goldberg, E., *The New Executive Brain: Frontal Lobes in a Complex World*, Oxford University Press, 2009.

[36] Goldberg, D.E., *Genetic Algorithms in Search, Optimization and Machine Learning*, Addison Wesley, 1989.

[37] Goss, S., Aron, S., Deneubourg, J.L., and Pasteels, J.M.: "Self-organized shortcuts in the argentine ant," *Naturwissenschaften*, vol. 76, pp. 579–581, 1989.

[38] Gracia-Lázaro, C., Ferrer, A., Ruiz, G., Tarancón, A., Cuesta, J.A., Sánchez, A., and Morenoa, Y., "Heterogeneous networks do not promote cooperation when humans play a Prisoner's Dilemma," *Proceedings of the National Academy of Sciences*, PNAS, vol.109, no.32, pp.12922–1292, 2012.

[39] Grant, P.R., and Grant, B.R., *How and Why Species Multiply: The Radiation of Darwin's Finches*, Princeton Series in Evolutionary Biology, Princeton Univ Press, 2011.

[40] Gravner, J., and Griffeath, D., "Modeling snow crystal growth II: A mesoscopic lattice map with plausible dynamics," *Physica D: Nonlinear Phenomena*, vol.237, no 3, pp.385–404, 2008.

[41] Gravner, J., and Griffeath, D., "Modeling snow-crystal growth: A three- dimensional mesoscopic approach," *Physical Review E*, vol.79, 2009.

[42] Gould, S.J., *Bully for Brontosaurus: Reflections in Natural History*, W.W.Norton & Co.Inc., 1992.

[43] Haidt, J., *The Righteous Mind: Why Good People Are Divided by Politics and Religion*, Vintage, 2013.

[44] Harman, O., *The Price of Altruism: George Price and the Search for the Origins of Kindness*, W.W.Norton & Co.Inc., 2011.

[45] He, C., Noman, N., and Iba, H., "An Improved Artificial Bee Colony Algorithm with Non-separable Operator," in *Proc. of International Conference on Convergence and Hybrid Information Technology*, 2012.

[46] Hepper, F., and Grenader, U., "A stochastic nonlinear model for coordinated bird flocks," AAAS publication, Washington,DC, 1990.

[47] Herdy, M., "Appication of the Evolution Strategy to Discrete Optimization Problems," in Parallel Problem Solving from Nature (PPSN), Schwefel, H.-P. and Männer, R. (eds.), pp.188–192, Springer-Verlag, 1990.

[48] Higashi, N., and Iba, H.: "Particle Swarm optimization with gaussian mutation," *Proceedings of IEEE Swarm Intelligence Symposium (SIS03)*, pp.72–79, 2003.

[49] Hilditch, C.J., Linear Skeletons From Square Cupboards in Meltzer,B. & Michie,D. (eds.), *Machine Intelligence 4*, pp.403–420, Edinburgh University Press, 1969.

[50] Hoel, E.P., Albantakis, L., and Tononi, G., "Quantifying causal emergence shows that macro can beat micro," *Proceedings of the National Academy of Sciences*, PNAS, vol.110, no.49, pp.19790–19795, 2013.

[51] Holland, J.H.: *Adaptation in Natural and Artificial Systems*. University of Michigan press, 1975.

[52] Huberman, B.A., and Glance, N.S.: "Evolutionary games and computer simulations," *Proceedings of the National Academy of Sciences*, PNAS, vol.90, no.16, pp.7716–7718, 1993.

[53] Iba, H., and Noman, N.: "New Frontiers in Evolutionary Algorithms: Theory and Applications," ISBN-10:1848166818, World Scientific Publishing Company, 2011.

[54] Iba, H.: "Evolutionary Approach to Machine Learning and Deep Neural Networks: Neuro-Evolution and Gene," ISBN-10:9811301999, Springer, 2018.

[55] Ichinose, G., and Masuda, N., "Zero-determinant strategies in finitely repeated games," *Journal of Theoretical Biology*, vol.438, pp.61–77, 2018.

[56] Inglett, J.E., and Rodriguez-Seda, E.J., "Object Transportation by Cooperative Robots," in *Proc. SoutheastCon 2017*, Charlotte, NC, pp. 1-6, 2017.

[57] Ishibuchi, H., Tsukamoto, N., and Nojima, Y., "Evolutionary many-objective optimization: A short review," in *Proc. IEEE Congress on Evolutionary Computation*, pp. 2419–2426, 2008.

[58] Karaboga, D., and Basturk, B.: "A powerful and efficient algorithm for numerical function optimization: Artificial Bee Colony (ABC) algorithm," *Journal of Global Optimization*, vol.39, pp.459–471, 2007.

[59] Karaboga, D., Gorkemli, B., Ozturk. C., and Karaboga, N.: "A Comprehensive Survey: Artificial Bee Colony (ABC) Algorithm and Applications," *Artificial Intelligence Review*, Doi:10.1007/s10462-012-9328-0, 2012.

[60] Kendall, G., Yao, X., and Chong, S.-Y.: "The Iterated Prisoners' Dilemma: 20 Years on," World Scientific Pub Co Inc., 2007.

[61] Kenrick, D.T., *Sex Murder and the Meaning of Life: A Psychologist Investigates How Evolution Cognition and Complexity are Revolutionizing our View of Human Nature*, Basic Books, 2011.

[62] Kennedy, J., and Eberhart, R.C.: "Particle swarm optimization," *Proceedings of IEEE the International Conference on Neural Networks*, pp.1942–1948, 1995.

[63] Kennedy, J., and Eberhart, R.C.: "Swarm Intelligence," Morgan Kaufmann Publishers, 2001.

[64] Kondo, S., and Asai, R., "A reaction diffusion wave on the skin of the marine angelfish Pomacanthus," *Nature*, vol.376, pp.765-?768, 1995.

[65] Kusch, I., and Markus, M.: "Mollusc Shell pigmentation: Cellular automaton simulations and evidence for undecidability," *Journal of theoretical biology*, vol.178, pp.333-340, 1996.

[66] Landaez, Y., Dohi, H., and Iba, H., "Swarm Intelligence for Object Retrieval Applying Cooperative Transportation in Unknown Environments," in *Proc. 2018 4th International Conference on Robotics and Artificial Intelligence (ICRAI 2018)*, 2018.

[67] Levesque, H.J., "Is it enough to get the behaviour right?," *Proc. of 21st Joint Conference on Artificial Intelligence (IJCAI-09)*, pp.1439–1444, 2009.

[68] Levitt, S.D., and Dubner, S.J., *Freakonomics – A Rogue Economist Explores The Hidden Side Of Everything*, William Morrow/Harper-collin, 2005.

[69] Li, A., and Yong, X., "Entanglement Guarantees Emergence of Co-operation in Quantum Prisoner's Dilemma Games on Networks," doi:10.1038/srep06286 Scientific Reports 4, Article number: 6286, 2014.

[70] Lohmann, R., "Structure evolution and incomplete induction," in *Proc. 2nd Parallel Problem Solving from Nature (PPSN92)*, pp.175–185, North-Holland, 1992.

[71] Martinez, G.J.: "Introduction to Rule 110," Rule 110 Winter WorkShop, 2004 http://www.rule110.org/amhso/results/rule110-intro/introRule110.html

[72] Melo, R.S., Macharet, D.G., and Campos, M.F.M., "Multi–object Transportation Using a Mobile Robot," in *Proc. 12th Latin American Robotics Symposium and 3rd Brazilian Symposium on Robotics (LARS-SBR)*, pp.234–239, 2015.

[73] Mesoudi, A., *Cultural Evolution: How Darwinian Theory Can Explain Human Culture and Synthesize the Social Sciences*, University of Chicago Press, 2011.

[74] Michalewics, Z., *Genetic Algorithms + Data Structures = Evolution Programs*, Springer-Verlag, 1992.

[75] Mitchell, M.: "Life and evolution in computers," *History and Philosophy of the Life Sciences* vol.23, pp.361–383, 2001.

[76] Mitchell, M.: "Complexity: A Guided Tour," Oxford University Press, 2009.

[77] Mochizuki, A., "Pattern formation of the cone mosaic in the zebrafish retina: a cell rearrangement model," *J. Theor. Biol.*, vol.215, no.3, pp.345–61, 2002.

[78] Murray, J.D., Mathematical Biology: I. An Introduction, Springer, 3rd printing 2008.

[79] Murray, J.D., Mathematical Biology II: Spatial Models and Biomedical Applications, Springer, 3rd printing 2008.

[80] Nagel, K., and Shreckenberg, M.: "A cellular automaton model for freeway traffic," *Journal de Physique I*, vol.2, no.12, pp.2221–2229, 1992.

[81] Nakagaki, T., Yamada, H., and Toth, A., "Intelligence: Maze-Solving by an Amoeboid Organism," *Nature*, vol.407, p.470, 2000.

[82] Neary, T., and Woods, D.: "P-completeness of cellular automaton Rule 110," *Proceedings of ICALP 2006 - International Colloquium on Automata Languages and Programming*, Lecture Notes in Computer Science, vol.4051, pp.132-143, Springer, 2006.

[83] Ninagawa, S.: "1/f noise in elementary cellular automaton rule 110," *Proceedings of the 5th international conference on Unconventional Computation, UC06*, Lecture Notes in Computer Science, vol.4135/2006, pp.207–216, Springer, 2006.

[84] Nolfi, S., and Floreano, D. *Evolutionary Robotics*, MIT Press, 2000.

[85] Nowak, M.A.: "Evolutionary Dynamics: Exploring the Equations of Life," Belknap Press of Harvard University Press , 2006.

[86] Onuma, H., Okubo, A., Yokokawa, M., Endo, M., Kurihashi, A., and Sawahata, H., "Rebirth of a Dead Belousov–Zhabotinsky Oscillator," *The Journal of Physical Chemistry A*, vol.115, no.49, pp14137–14142, 2011.

[87] Parkinson, R., "The Dvorak Simplified Keyboard: Forty years of Frustration," *Computers and Automation magazine*, vol.21, pp.18–25, November, 1972.

[88] Penrose, R., *The Emperor's New Mind: Concerning Computers, Minds, and The Laws of Physics*, Pxford Univ Press, 1989.

[89] Penrose, R., "Beyond the doubting of a shadow a reply to commentaries on shadows of the mind," *PSYCHE: An Interdisciplinary Journal of Research On Consciousness*, vol.2, no.23, 1996.

[90] Pinker, S., *How the Mind Works*, W.W.Norton & Co.Inc., 1991.

[91] Pinker, S., *The Blank Slate: The Modern Denial of Human Nature*, Penguin Books, 2003.

[92] Pinker, S., *The Language Instinct: How The Mind Creates Language*, Harper Perennial Modern Classics, 2007.

[93] Pinker, S., *The Better Angels of Our Nature: Why Violence Has Declined*, Penguin Books, 2012.

[94] Pinto, A., Oates, J., Grutter, A., and Bshary, R., "Cleaner wrasses Labroides dimidiatus are more cooperative in the presence of an audience," *Current Biology*, vol.21, no.13, pp.1140–1144, 2011.

[95] Press, W.H., and Dyson, F.J., "Iterated Prisoner's Dilemma contains strategies that dominate any evolutionary opponent," *PNAS: Proceedings of the National Academy of Sciences*, vol.109, no.26, pp.10409–10413, 2012.

[96] Rajewsky, N., Santen, L., Schadschneider, A., and Schreckenberg, M.: "The asymmetric exclusion process: Comparison of update procedures," *Journal of Statistical Physics*, vol.92, pp.151–194, 1998.

[97] Rechenberg, I., "Evolution strategy and human decision making," *Human decision making and manual control*, Willumeit, H. P. (ed.), pp.349–359, North-Holland, 1986.

[98] Reynolds, C.W.: "Flocks, herds and schools: A distributed behavioral model," *Computer Graphics*, vol.21, no.4, pp.25–34, 1987.

[99] Ridley, M., *The Agile Gene: How Nature Turns on Nurture*, Harper Perennial, 2004.

[100] Ridley, M., "The Evolution of Everything: How New Ideas Emerge," Harper, 2015.

[101] Robbins, C., and Morrell, J., *Mold, Housing and Wood*, Western Wood Products Association, 2002.

[102] Rucker, R.: "Artificial Life Lab," Waite Group Press, 1993.

[103] Sadhu, A.K., Rakshit, P., and Konar, A., "A modified Imperialist Competitive Algorithm for multi-robot stick-carrying application," *Robotics and Autonomous Systems*, vol.76, pp.15–35, 2015.

[104] Sandel, M.: "Justice: What's the Right Thing to Do?" Penguin, 2010.

[105] Schaffer, J.D., and Grefenstette, J.J., "Multi-Objective Learning via Genetic Algorithms," in *Proc. of the 9th International Joint Conference on Artificial Intelligence*, pp.593-595, 1985.

[106] Schaffer, J.D., "Multiple Objective Optimization with Vector Evaluated Genetic Algorithms," in *Proc. of an International Conference on Genetic Algorithms and Their Applications*, pp.93-100, 1985.

[107] Schelling, T.C.: "Dynamic models of segregation," *Journal of Mathematical Sociology*, vol.1., pp.143-186, 1971.

[108] Schwefel, H., *Numerical optimization of computer models*, John Wiley & Sons, 1981.

[109] Segerstrale, U., *Defenders of the Truth: The Sociobiology Debate*, Oxford Univ Press, 2001.

[110] Shayestegan, M., and Marhaban, M.H., "A Braitenberg Approach to Mobile Robot Navigation in Unknown Environments," in Ponnambalam, S.G., Parkkinen, J., Ramanathan, K.C. (eds.), *Trends in Intelligent Robotics, Automation, and Manufacturing. IRAM 2012*, Communications in Computer and Information Science, vol 330. Springer, 2012.

[111] Smith, J.M., and Szathmary, E., "The Origins of Life: From the Birth of Life to the Origin of Language," Oxford University Press, 2000.

[112] Sörensen, K., "Metaheuristics–the metaphor exposed," *International Transactions in Operational Research*, vol.22, no.1, pp.3–18, 2015.

[113] Sörensen, K., Sevaux, M., and Glover, F., "A History of Metaheuristics," arXiv:1704.00853v1 [cs.AI] 4 Apr 2017, to appear in Mart,R., Pardalos,P., and Resende,M., *Handbook of Heuristics*, Springer.

[114] Stewart, I., *The Mathematics of Life*, Basic Books, 2013.

[115] Tamura, H., "A comparison of line thinning algorithms from digital geometry viewpoint," in *Proc. Int. Joint Conf. on Pattern Recognition*, Kyoto, Japan, pp.715–719, 1978.

[116] Tero, A., Takagi, S., Saigusa, T., Ito, K., Bebber, D.P., Fricker, M.D., Yumiki, K., Kobayashi, R., and Nakagaki, T., "Rules for Biologically Inspired Adaptive Network Design," *Science*, vol.327, no.5964, pp.439–442, 2010.

[117] Torabi, S., "Collective Transportation of Objects by a Swarm of Robots," Master Thesis in Complex Adaptative Systems, Chalmers University of Technology, 2015.

[118] Uchiyama, N., Mori, A., Kajita, Y., Sano, S., and Takagi, S., "Object-transportation control for a human-operated robotic manipulator," in *Proc. 2008 IEEE International Conference on Emerging Technologies and Factory Automation*, pp.164–169, 2008.

[119] Vukov, J., Szabó, G., and Szolnoki, A., "Evolutionary prisoner's dilemma game on Newman-Watts networks," Phys. Rev.E, vol.77, pp.026109–, 2008.

[120] Wang, Y., and de Silva, C.W., "An object transportation system with multiple robots and machine learning," in *Proc. of the 2005 American Control Conference*, pp.1371–1376, 2005.

[121] Weyland, D., "A critical analysis of the harmony search algorithm? How not to solve sudoku," *Operations Research Perspectives*, vol.2, pp.97–105, 2015.

[122] Weyland, D., "A Rigorous Analysis of the Harmony Search Algorithm - How the Research Community can be misled by a "novel" Methodology," *International Journal of Applied Metaheuristic Computing*, vol.1, no.2, pp.50–60, 2010.

[123] Wilson, E.O., *The Insect Societies*, Belknap Press of Harvard University Press, 1971.

[124] Wilson, E.O., *Sociobiology: The Abridged Edition Abridged Edition*, Belknap Press of Harvard University Press, 1980.

[125] Witten, T.A., and Sander, L.M., "Diffusion-limited aggregation, a kinetic critical phenomenon," *Phys. Rev. Lett.*, vol.47, pp.1400-1403, 1981.

[126] Wolfram, S.: "A New Kind of Science," Wolfram Media, 2002.

[127] Yamashita, A., Fukuchi, M., Ota, J., Arai, T., and Asama, H., "Motion planning for cooperative transportation of a large object by multiple mobile robots in a 3D environment," in *Proc. 2000 ICRA Millennium Conference, IEEE International Conference on Robotics and Automation*, pp.3144–3151, 2000.

[128] Yang, X.-S., and Deb, S., 2009. "Cuckoo search via Levy flights," in *Proc. World Congress on Nature & Biologically Inspired Computing (NaBIC 2009)*, pp.210–214, IEEE Publications, 2009.

[129] Yang, X., *Nature-Inspired Metaheuristic Algorithms*, 2nd ed., Luniver Press, 2010.

Index

Color Section

Chapter 2: Fig. 2.19, p. 41

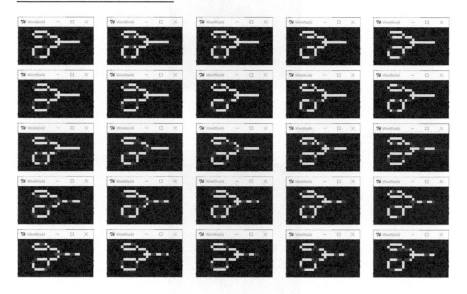

Chapter 4: Fig. 4.14, p. 108

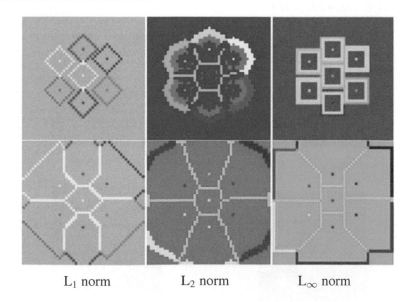

L_1 norm $\quad\quad\quad$ L_2 norm $\quad\quad\quad$ L_∞ norm

Chapter 4: Fig. 4.29, p. 129

Chapter 4: Fig. 4.30, p. 129

Chapter 4: Fig. 4.37, p. 134

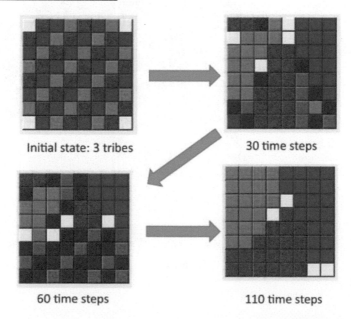

Initial state: 3 tribes

30 time steps

60 time steps

110 time steps

Chapter 5: Fig. 5.18, p. 155

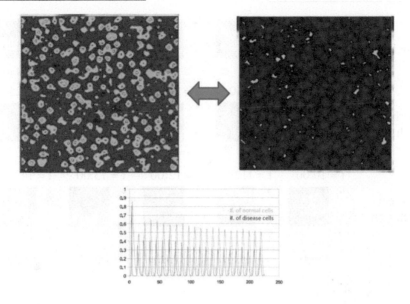

Chapter 6: Fig. 6.5, p. 182

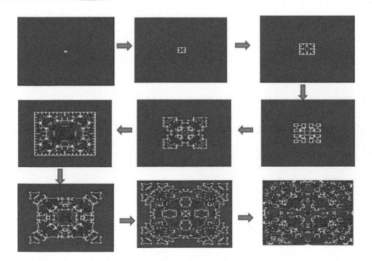

Chapter 6: Fig. 6.29, p. 207

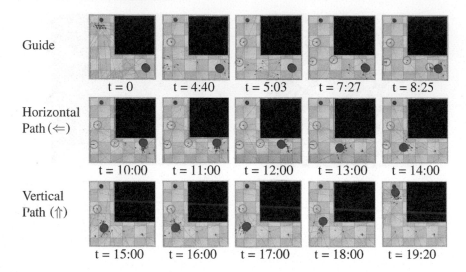

Guide   t = 0   t = 4:40   t = 5:03   t = 7:27   t = 8:25

Horizontal Path (⇐)   t = 10:00   t = 11:00   t = 12:00   t = 13:00   t = 14:00

Vertical Path (⇑)   t = 15:00   t = 16:00   t = 17:00   t = 18:00   t = 19:20